Urban Sustainability

Editor-in-Chief

Ali Cheshmehzangi, School of Architecture, Design and Planning, The University of Queensland, Brisbane, QLD, Australia

The Urban Sustainability Book Series is a valuable resource for sustainability and urban-related education and research. It offers an inter-disciplinary platform covering all four areas of practice, policy, education, research, and their nexus. The publications in this series are related to critical areas of sustainability, urban studies, planning, and urban geography.

This book series aims to put together cutting-edge research findings linked to the overarching field of urban sustainability. The scope and nature of the topic are broad and interdisciplinary and bring together various associated disciplines from sustainable development, environmental sciences, urbanism, etc. With many advanced research findings in the field, there is a need to put together various discussions and contributions on specific sustainability fields, covering a good range of topics on sustainable development, sustainable urbanism, and urban sustainability. Despite the broad range of issues, we note the importance of practical and policy-oriented directions, extending the literature and directions and pathways towards achieving urban sustainability.

The series will appeal to urbanists, geographers, planners, engineers, architects, governmental authorities, policymakers, researchers of all levels, and to all of those interested in a wide-ranging overview of urban sustainability and its associated fields. The series includes monographs and edited volumes, covering a range of topics under the urban sustainability topic, which can also be used for teaching materials.

Martin de Jong · Daan Schraven · Tong Xin ·
Liang Dong
Editors

The Inclusive Circular Economy

Challenges and Opportunities for Urban Innovation

Editors
Martin de Jong
Rotterdam School of Management &
Erasmus School of Law
Erasmus University Rotterdam
Rotterdam, The Netherlands

Smart City Institute, HEC Liege
Liege, Belgium

Institute of Global Public Policy
Fudan University
Shanghai, China

Tong Xin
College of Urban and Environmental
Sciences
Peking University
Beijing, China

Daan Schraven
Faculty of Architecture in the Built
Environment
Delft University of Technology
Delft, The Netherlands

Liang Dong
City University of Hong Kong
Kowloon, Hong Kong

ISSN 2731-6483 ISSN 2731-6491 (electronic)
Urban Sustainability
ISBN 978-981-96-6866-3 ISBN 978-981-96-6867-0 (eBook)
https://doi.org/10.1007/978-981-96-6867-0

© The Editor(s) (if applicable) and The Author(s) 2026. This book is an open access publication.

Open Access This book is licensed under the terms of the Creative Commons Attribution 4.0 International License (http://creativecommons.org/licenses/by/4.0/), which permits use, sharing, adaptation, distribution and reproduction in any medium or format, as long as you give appropriate credit to the original author(s) and the source, provide a link to the Creative Commons license and indicate if changes were made.
The images or other third party material in this book are included in the book's Creative Commons license, unless indicated otherwise in a credit line to the material. If material is not included in the book's Creative Commons license and your intended use is not permitted by statutory regulation or exceeds the permitted use, you will need to obtain permission directly from the copyright holder.
The use of general descriptive names, registered names, trademarks, service marks, etc. in this publication does not imply, even in the absence of a specific statement, that such names are exempt from the relevant protective laws and regulations and therefore free for general use.
The publisher, the authors and the editors are safe to assume that the advice and information in this book are believed to be true and accurate at the date of publication. Neither the publisher nor the authors or the editors give a warranty, expressed or implied, with respect to the material contained herein or for any errors or omissions that may have been made. The publisher remains neutral with regard to jurisdictional claims in published maps and institutional affiliations.

This Springer imprint is published by the registered company Springer Nature Singapore Pte Ltd.
The registered company address is: 152 Beach Road, #21-01/04 Gateway East, Singapore 189721, Singapore

If disposing of this product, please recycle the paper.

Contents

Introduction 1
Martin de Jong, Daan Schraven, Anne Hofmann, and Liang Dong

Exploring the Circular City: A Bibliometric and Definition
Analysis 11
Quirien Reijtenbagh, Zhaowen Liu, and Daan Schraven

Exploring the Inclusive City: Definitions and Dimensions 41
Danni Liang, Martin de Jong, and Daan Schraven

A Conceptual Framework for Inclusive Circular Urban Waste
Management Systems 69
Daan Schraven, Filippos K. Zisopoulos, Liang Dong, and Martin de Jong

Managing the Transition to a Circular Urban Waste Management
System 91
Afsaneh Moradi

Accumulation of Circular Economy Policy in China: Goals,
Instruments and Demonstration Projects Over 2006–2022 113
Wenting Ma, Martin de Jong, and Thomas Hoppe

Building the Zero Waste City: A Half Century of Efforts in Beijing 135
Xin Tong

Household Renovation Waste in the Netherlands: Mapping
the Social Side of Waste Flows 149
Daan Schraven, Kai Vaessen, Zhaowen Liu, and Tong Wang

Business Model Innovations in Post-Consumer Recycling in Urban
China 179
Xin Tong

Informal Reuse, Repair and Refurbishment Business Networks for Air Conditioners in Gangxia Village, Shenzhen 195
Yuk Tung Chow and Benjamin Steuer

Conclusions 213
Daan Schraven, Martin de Jong, Zhaowen Liu, and Xin Tong

About the Editors

Dr. Liang Dong is an Associate Professor in the Department of Public and International Affairs (PIA), and School of Energy and Environment (SEE), City University of Hong Kong. He focuses to interdisciplinary and international collaborative research on urban sustainability, sustainability science & policy, and their application in circular economy, corporate environmental-social-governance (ESG), Net Zero and carbon neutrality policy studies, sustainable human-techno-environmental system design, and sustainable urban planning, using advanced life cycle based, spatial and data driven techniques. He served as the committee member of the "Industrial Symbiosis" section of the International Society for Industrial Ecology, Industrial Ecology section of Ecological Society of China, Circular economy section of Chinese Society for Environmental Sciences, and Vice Chairman of the Urban Human Settlement Environment Committee of the Chinese Society For Urban Studies. His research is well recognised by international society and was awarded as the Stanford University World Top 2% most-cited Scientists (2020 to 2025).

Prof. Martin de Jong is a full professor and scientific director of the Erasmus Initiative for the Dynamics of Inclusive Prosperity at ErasmusUniversity Rotterdam. He is also professor at Rotterdam School of Management and Erasmus School of Law. He holds visiting professor positions at the Institute of Global Public Policy, Fudan University (Shanghai, China) and the Smart City Institute of HEC-Liège, Université de Liège (Belgium). He publishes and lectures extensively on topics of sustainable urban development, city branding, policy-making and China studies.

Dr. Daan Schraven is an associate professor on new economics in the built environment at Delft University of Technology, affiliated to the Faculty of Architecture in the Built Environment. He is a business economist by background with an empirical affinity to matters related to the built environment. In this capacity his research interests include the economic implications of the circular transition in the built environment, new economic logics for a sustainable urban future of the city, and the value and valuation of assets and materials in the urban space, markets and supply chains. Daan is principal investigator to various industry-funded projects and (inter)national science foundation grants where he uses research by design approaches closely with practice to explore and develop new business models and valuation methods to facilitate societal transitions toward more sustainability and circularity. He is a board member of the Centre for Sustainability, on the theme circular cities and regions, and member of the Domain Acceleration Team on Sustainability & Circularity of the four Dutch Universities of Technology (4TU).

Prof. Tong Xin is an associate professor at Peking University. Prof. Tong has been working on extended producer responsibility in China's e-waste management for a number of years, specifically focusing on the interactions between global environmental governance and technological change in developing countries. She has been active in consultancy positions for local governments in Beijing, Sichuan, Shenzhen, Ningbo, and many other large Chinese cities. Prof. Tong is currently involved in two national research and development programs on solid waste management focused on behavioral change at a community level and supporting infrastructure with ICT applications.

List of Figures

Introduction

Fig. 1	Schema of urban waste management system, adapted from Liu et al. [5]	4
Fig. 2	Structure of the book	5

Exploring the Circular City: A Bibliometric and Definition Analysis

Fig. 1	Average Publication year versus number of publications per keyword	15
Fig. 2	Boxplot spread of Publication year sequence per keyword	15
Fig. 3	**a.** Social network analysis. **b.** Cluster analysis	19
Fig. 4	A conceptual framework of the circular city	26

Exploring the Inclusive City: Definitions and Dimensions

Fig. 1	Research design	50
Fig. 2	Total number of publications about inclusive cities research (2000–2022)	51
Fig. 3	Disciplines and their publication numbers regarding the inclusive city (2000–2022)	51
Fig. 4	Co-occurrence network of high-frequency keywords	53
Fig. 5	The tree diagram of cluster analysis	54
Fig. 6	Changes in the dimensions and connotations of the inclusive city	63

A Conceptual Framework for Inclusive Circular Urban Waste Management Systems

Fig. 1	Setup following the double diamond design-based research method	71
Fig. 2	Connections between description of challenges and necessity for framework	78

Fig. 3	Methodology for developing inclusive and circular UWMSs	78
Fig. 4	System boundaries, actors, resource flows, and indicators useful for the development of inclusive and circular UWMSs	81
Fig. 5	Final design	82

Managing the Transition to a Circular Urban Waste Management System

| Fig. 1 | Timeline for the transition to a circular economy during the period of 2016 to 2050. Developed based on the data from [16] | 95 |

Building the Zero Waste City: A Half Century of Efforts in Beijing

| Fig. 1 | The spatial flows of waste across regions | 138 |
| Fig. 2 | The social network supporting the informal recycling sector | 140 |

Household Renovation Waste in the Netherlands: Mapping the Social Side of Waste Flows

Fig. 1	Schematic representation of theory of planned behavior with examples inspired by Pongpunpurt et al. [39]	155
Fig. 2	Waste Journey conceptualization inspired by Customer Journey	157
Fig. 3	Graphical representation of Waste Journey setup	159
Fig. 4	HR waste journey stages derived from theory	163
Fig. 5	Observed behaviors in Waste Journey 1	166
Fig. 6	Observed behaviors in Waste Journey 2	169

Business Model Innovations in Post-Consumer Recycling in Urban China

| Fig. 1 | IT solution as the center of the new business model | 184 |

Informal Reuse, Repair and Refurbishment Business Networks for Air Conditioners in Gangxia Village, Shenzhen

Fig. 1	Officially documented quantities of recovered WEEE in China (based on [18])	199
Fig. 2	Institutional evolution of China's WEEE management system (issuance dates in grey indicate amended versions)	201
Fig. 3	National regulations on WEEE management with particular focus on reuse, repair, refurbishment and remanufacturing (the author based on www.lawinfochina.com)	202
Fig. 4	Subsidies (CNY/unit) for formal WEEE processing companies (based on MOF et al. [19–21])	203

Fig. 5	Central government and Shenzhen's formalisation of WEEE management along the R-principles	205

Conclusions

Fig. 1	Layers of an UWMS	215
Fig. 2	Mapping of chapter relationships throughout the book	219

List of Tables

Exploring the Circular City: A Bibliometric and Definition Analysis

Table 1	Coding procedure of circular city definitions	17
Table 2	Occurrence analysis of most associated keywords to circular city	18
Table 3	Percentage diverging/converging keywords	20
Table 4	Results on aspects of the circular city that emerged from the coding process	21
Table 5	Occurrence of categories and sub-categories in definitions	23

Exploring the Inclusive City: Definitions and Dimensions

Table 1	Overview of selected books, book chapters and reports for qualitative review	45
Table 2	Complete list of merged and renamed keywords	48
Table 3	High-frequency keywords related to the inclusive city	52
Table 4	Comparison of the clustering results of the two studies	55
Table 5	Key terms of each dimension from all authors	61

A Conceptual Framework for Inclusive Circular Urban Waste Management Systems

Table 1	Challenges and required functions for developing the framework	77

Managing the Transition to a Circular Urban Waste Management System

Table 1	Waste circular-knowledge center circular economy [39]	106

Accumulation of Circular Economy Policy in China: Goals, Instruments and Demonstration Projects Over 2006–2022

Table 1	Overview of the key environmental governance strategies in China (2003–2022)	119
Table 2	CE demonstration projects by NDRC and MEE (see Appendix for full list of participating provinces, cities, districts and counties)	126
Table A1	Number of CE policy instruments over 2006–2021, classified as per policy instrument type [20]	129
Table A2	Three Chinese National circular economy city programs & related pilot initiatives	130

Household Renovation Waste in the Netherlands: Mapping the Social Side of Waste Flows

Table 1	Overview of HRW actors	164
Table 2	Overview interviewees of WJ1 about direct municipal offering	165
Table 3	Overview interviews of WJ2 about on-site container	168

Informal Reuse, Repair and Refurbishment Business Networks for Air Conditioners in Gangxia Village, Shenzhen

Table 1	Survey questions asked to respondents in the WEEE refurbishing and reuse sector in Gangxia village, Shenzhen	197
Table 2	Division of labour along the product value-chain for AC repair, refurbishment and reselling practices by informal entrepreneurs	206
Table 3	The operation costs of AC repair and resell shops	207

Introduction

Martin de Jong, Daan Schraven, Anne Hofmann, and Liang Dong

Abstract This chapter provides the background of this book's topic. It does so by explaining how informal waste picking and urban formal waste infrastructure systems are both relevant to realizing an effective inclusive and circular economy at the urban scale, but that bringing them together does not occur automatically in the policymaking process. It provides a conceptual model that clarifies how the different components of the urban waste management system are connected, clarifies the logic underlying the structuring of the book into the various chapters that follow and then proceeds to present a brief outline of what each of those following chapters will be dealing with.

Keywords Informal waste pickers · Circularity · Inclusion · Urban waste infrastructure management system · Overview of the chapters

---

M. de Jong (✉) · A. Hofmann
Rotterdam School of Management & Erasmus School of Law, Erasmus University Rotterdam, Rotterdam, The Netherlands
e-mail: w.m.jong@law.eur.nl; w.m.dejong@rsm.nl

A. Hofmann
e-mail: anne@jus.hofmann.de

M. de Jong
HEC-Liege, University of Liege, Liege, Belgium

Institute of Global Public Policy, Fudan University, Shanghai, China

D. Schraven
Faculty of Architecture in the Built Environment, Delft University of Technology, Delft, The Netherlands
e-mail: d.f.j.schraven@tudelft.nl

A. Hofmann
Faculty of Applied Sciences, Delft University of Technology, Delft, The Netherlands

L. Dong
City University of Hong Kong, Kowloon, Hong Kong
e-mail: liadong@cityu.edu.hk

© The Author(s) 2026
M. de Jong et al. (eds.), *The Inclusive Circular Economy*, Urban Sustainability,
https://doi.org/10.1007/978-981-96-6867-0_1

Whoever is attentive and attuned to the phenomena can sometimes spot them: unknown and somewhat shy individuals operating nearby public litter bins and garbage containers scavenging for useful elements they expect can be collected and sold at a reasonable price. We know them as informal waste pickers. In the larger cities of most developing countries, they are in fact quite common and contribute significantly to the reuse and recycling of materials that would otherwise end up incinerated or landfilled [3]. For example, informal waste pickers were found to have a crucial role in Accra (Ghana) and Porto Alegre (Brazil) in the implementation of waste policies in these cities [1]. In wealthier countries they have been mostly absent for decades but represent something reemerging now that urban migrants and underclasses have lost purchasing power year after year and have become reliant on it. However, the academic debate around informal waste pickers in wealthier countries focuses on discrimination, stigma, and social exclusion, and less on their environmental contributions or how they can be included in the formal waste management processes [7].

The role of informal waste pickers is set to become a lot bigger due to a few recent trends. First of all, the overall population of cities is growing. In 2020, already 55% of the world's population lived in cities, and this proportion is expected to increase to 68% by 2050 [2]. And as waste generating enterprises, cities harbor a diverse set of people with different urban lifestyles, which is likely to increase the amount and complexity of municipal waste [4, 8].

Second, both emerging and mature economies are busy introducing policies to realize circular economies for themselves but handle those transitions in different ways. In emerging economies, governments recognize the existence of the informal waste picker. However, they choose to implement policies to strengthen waste collection and processing rather through official channels. For example, China has introduced quite a few policies and formal enforcement mechanisms through legislation and market regulation for e-waste, creating a further disconnect with the informal recycling sector [9]. This goes potentially at the expense of the informal sector although there also appear opportunities to integrate them into a more complete and comprehensive waste management system.

In economically mature countries in Europe, such as the Netherlands, where waste collection systems are more advanced, their recent appearance makes most citizens somewhat uncomfortable but their market opportunities as such appear satisfactory as their growing numbers are showing. For example, legislation affecting the circular economy is enacted including extended producer responsibility banning single use plastics, mandating labelling on plastic products and promoting waste sorting and public awareness [6]. This helped to introduce deposits for cans and plastic bottles as a refund system, which inspired people to start collecting these bottles and bring them to the supermarket. In these ways they may well constitute a major underestimated asset to the future of the circular economy.

Both urbanization and the transition to a circular economy instill the need to find effective waste management policies and their effective implementation into sustainable circular practices of production and consumption. It evokes the question how the inclusion of informal waste collection activities be organized and made part of an integrated and well-functioning circular economy? The academic literature in urban and environmental studies of recent years pays ample attention to the mapping, measurement and implementation of policies promoting the circular economy. In the social and behavioral sciences, the relative importance of conceptualizing and realizing various types of inclusion has shown an enormous increase. What is far less common is to build the analytical connection between these two developments although the value of that has become more and more apparent.

The value seems to lie in the unification of the goals that both movements toward circularity and inclusion have. On the one hand environmental and spatial aspects seem to be well-connected with each other, sometimes even unified into one perspective. Closing the circle from delving resources from the natural environment through manufacturing, sales, consumption and disposal decreases the amounts of waste dumped into urban and rural space and reinserts potentially valuable materials back into the production and consumption process and boosts spatial and environmental inclusion. On the other hand, unification can be considered far more indirect on the various aspects of social, political and economic inclusion. These inclusion aspects give vulnerable groups a say in decision-making as to how the chain of handling materials from cradle to grave is organized, which is also clearly of major value to realizing higher levels of circularity in the economy.

One could ask under which type of economic system this unification can best be reached. Along the entire chain of exploitation, production, distribution, consumption and end of cycle, it is conceivable that enormous economic and ecological gains are obtained if circularity and inclusion are considered in combination. But in practice, things appear not that simple. On the one hand, the dominant capitalist practice of letting 'laissez-faire' market operations determine whether and where business opportunities emerge following a 'spontaneous order' tend to leave the identification of opportunities for inclusive circularity or circular inclusion to mere chance. On the other hand, heavy-handed government policies in which large-scale formal waste processing prevails over more detailed informal collection through marginal groups in society also overshoot their target: technology and engineering prowess alone are not always the answer to societal problem-solving either.

Is the city the appropriate focal point for change proposals? The urban level is many ways indeed the integrating scale for policy and practice. There are a few reasons that can be purported for this. First, waste generation is concentrated in the urban environment, as citizens generate waste in their households, and keep this in the urban vicinity. Second, at the city level, more comprehensive policies adopted at the national scale can be adopted, amended and implemented in their local context. Finally, the city level harbors the operational processes that require the intervention, for instance the infrastructure and logistics that are effectively established and governed at the urban and regional scale. Following these arguments would lead us to suggest that the city is a fruitful focal point to learn about an effective mesh of policies and practices for circularity and inclusion. As the city is suggested to the point of departure, the central question addressed in this book is raised as: *How should a city govern the entirety of its urban waste management system such that it achieves inclusive and circular aspirations?*

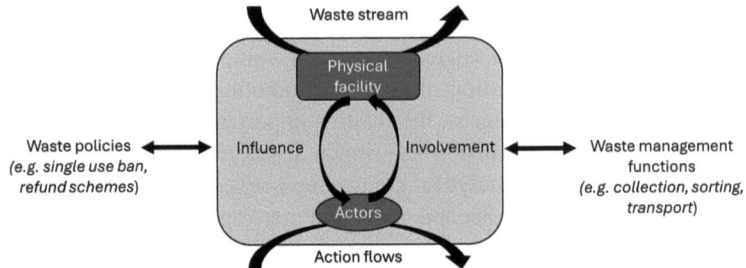

Fig. 1 Schema of urban waste management system, adapted from Liu et al. [5]

In essence, what is required is that waste management facilities are considered part of a broader systems perspective: individual elements such as (1) physical facilities, stocks and flows of materials, water and energy and (2) public, civil and private sector players that operate, own and control them as well as (3) the mutual relations these objects, infrastructures and urban and industrial actors have with each other need to be identified, categorized, mapped, structured and eventually reorganized.

It is only when developing the complete picture of how everything hangs together that possibilities to restructure elements and relations become fully visible and system performance can potentially be optimized. This is more easily said than done: knowing what a well-performing waste management system looks like is not tantamount to realizing it because one or more actors enjoying vested interests in the system and hampering change may have to be induced or forced to change. However, knowing at least is a first step to making transformation a future possibility.

This book approaches the topic of inclusive circular urban infrastructure systems. Liu et al. [5] described the building blocks of an urban waste management system from the municipal solid waste's perspective. Figure 1 shows a schematic representation of the urban waste management system. In the center there are two main blocks: physical facilities and actors who are interrelated in the process of managing waste. The public, civil and private actors are involved in (un)intended management functions, like collecting, sorting and transporting waste to a physical facility (e.g. sorting facility, recyclers), where the characteristics and laws (e.g. refund schemes or sorting requirements) that guide these facilities influence the ability of actors to perform those functions. The waste streams and actor activities serve as connectors to other physical facilities.

Following these building blocks of the urban waste management system, this book systematically examines what the academic literature has to say about circular urbanity, inclusive urbanity, synthesizes insights from both into a comprehensive framework enabling us to analyze waste infrastructure management systems at the urban scale. It subsequently offers a variety of empirical cases and examples from China and the Netherlands that illustrate what urban waste infrastructures look like, how they function, what options there are to improve their inclusiveness and levels of circularity, in which national contexts they are embedded and what lessons other cities can draw from them.

Introduction

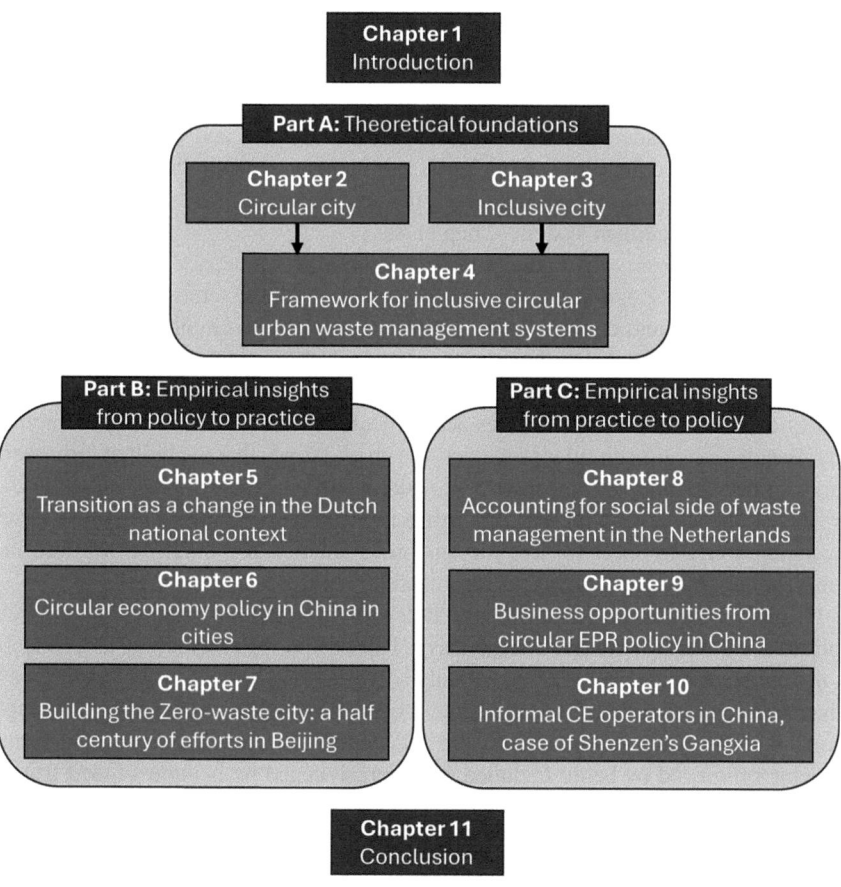

Fig. 2 Structure of the book

This overarching approach is structured in the book through three main parts (Fig. 2). The first part (A) addresses the theoretical foundations to develop an overarching framework for inclusive circular urban waste management systems. The following two (B and C) parts then use these theoretical underpinnings to discuss empirical insights from six cases on circular policy initiatives and their implications for waste management practices, and vice versa. The next sections narrate the particular points of attention in each of the following chapters and highlight some of the key take-aways.

1 Theoretical Foundations

Part A touches base on the theoretical foundations of circularity ambitions of a city, by studying the academic literature through bibliometrics and content review on the concepts of circular cities (Chapter 2) and inclusive cities (Chapter 3) and then designed a framework that encapsulates the requirements of an inclusive circular urban waste management system (Chapter 4).

Chapter 2 addresses the concept of the circular city. It departs as to why circularity is warranted: population is expected to grow in cities, setting expectations on increased amount and complexity of municipal waste. The circular economy is an alternative to the take-make-dispose economy by turning waste into resources. The chapter reviews how cities should become effectively circular as to when circular strategies are adopted. It contributes academically in terms of defining the circular city. Besides adopting circular strategies, the chapter finds that the built environment can also be strengthened in terms of its adaptation and inclusion of nature in the urban setting, suggesting the need for urban regeneration. Chapter 2 operationalizes the circular city by offering five dimensions, including a system description, the components, principles, goals and organization.

Chapter 3 reviews the inclusive city. It departs from the observation that municipalities often brand themselves as inclusive. Then it finds that inclusivity as a concept is not unique to any one field, but actually serves a great variety of inclusion types. Chapter 3 posits that inclusion interacts with its antonym, called exclusion, through which it becomes apparent that there are always two sides to coin for actors in terms of being privileged by being included or underprivileged by being excluded from a certain function. Chapter 3 proposes to look more anatomically into in what ways people are excluded from benefits, facilities or privileges. The chapter contributes by systematically examining what an inclusive city exactly entails, how it can be defined, the various dimensions that it has and how it can be achieved. The chapter presents a framework for an inclusive where it describes six dimensions of inclusion, i.e. spatial, social, environmental, economic, political and cultural inclusion.

To curb the high waste generation, Chapter 4 develops the conceptual framework by taking the schema of urban waste management systems and projecting aspirations of inclusion (from Chapter 3) and circularity (from Chapter 2) on it. Using the combination of the two views, Chapter 4 purports that a city actually needs to take more responsibility towards people and the natural ecosystem that surrounds the city in its pursuance of inclusion and circularity of waste management. This hints at an important implication in that waste management needs to be redefined as a resource collection and distribution network. In this way it is expected that a city can avoid a potential lock-in of its waste management infrastructure. This signifies a disability to make changes to the function of the system, like including local stakeholders in long term investments. The framework of chapter 4 is designed to include steps that help develop circularity principles whereby all stakeholders act responsibly and adhere to inclusive principles.

2 Empirical Insights from Policy to Practice

Part B makes a deep dive into three cases that describe empirical insights on the level where policy initiatives have implications for the waste management practices from the Netherlands and China. The chapters reveal how policies can direct changes in the Netherlands (Chapter 5) and China (Chapter 6). In addition, it also shows how effectuating policy changes becomes more difficult if the goals need to be upgraded by a fundamental scope expansion to better capture existing practices (Chapter 7).

Chapter 5 investigates the principles of transition management as they apply to changing an urban waste management system into a circular and more resilient one. It examines the different types of policies and practices in the Netherlands that have the potential to be used for a transition toward a circular waste management system. The chapter describes the history of waste management in the Netherlands in relation to its demographics in cities. It illustrates the need in the Dutch context for fundamental changes. Chapter 5 takes the transition management model as a systemic approach to analyze this long-term transformation of the Dutch waste management system. It thereby reviews the Waste Disposal Act of 1975 and the National Waste Management Plan of 1992 and recent developments that emerged in light of the circular economy transition.

Chapter 6 presents an analysis of the circular economy policies in China that were issued between 2006 and 2022 and shows how the government issued and implemented an increasing number of these over time. In particular, the 13th Five Year Plan (2016–2020) reveals a sudden increase in wide variety. It specifically notes the evolution of intensity from CE pilots to national demonstration cities to zero-waste cities. In content, Chapter 6 unveils why there was a shift in these policy attributes from a primary focus on production efficient to the adoption of a whole life cycle perspective, that spans across the supply chain. One of these reasons includes the rapid economic development and urban population growth between the 1980s and 2020s. Another reason turns out to be completing the connection of more restricted waste handling policies to the broader circular economy ambition which were already hinted at since the 1990s with synonyms such as cleaner production, industrial ecology and ecological modernization, but only appear in view more recently.

Chapter 7 presents a case study of Beijing as a long-time development of its urban waste management system with its parallel transformation of the informal recycling sector since marketization in late 1970s. It discusses the focus of policies to address environmental and resource challenges, yet these initially excluded the migrant scavengers from their local policy efforts. In essence, the challenge has become how the informal recycling sector needed to be integrated within the modernization of the existing urban waste management system. This case underscores the limitations of applying a merely linear approach from policy to practice and warrants a focus on empirical insights derived from the reverse relationship.

3 Empirical Insights from Practice to Policy

Part C then delves into three cases that describe empirical findings on the level where waste management practices have implications for circular policy making for the Netherlands and China. The chapters reveal how waste management practices can be better understood by following their social processes as waste is handled in daily practice (Chapter 8), and what the organic reaction to new opportunities is from the informal sector when new legislation is introduced to waste management practices (Chapter 9). Finally, Chapter 10 discusses how the achievement of circularity and inclusion can backfire if a policy does not allow the organic adaptation of waste management practice in case official state-led policies are put in place.

Chapter 8 argues that enhancing waste management practices requires analytical tools capable of observing the social dimension of those practices. The chapter proposes a method to include this dimension and then applies it to the waste management practices following two Dutch household renovation projects. The approach, called Waste Journey, offers a way to map the social processes on waste handling, showing where certain events in social processes can be ringfenced as areas for improvements. Chapter 8 contributes as a new tool a qualitative method to a field primarily dominated by quantitative applications. The chapter highlights the additional attention that the method can provide in terms of describing and revealing the role and impact of different actors as they interact with waste and shape the actual waste management practices. Particularly human reasons behind certain challenges regarding the inclusion and circularity of the waste management practices are laid bare through using this method.

Chapter 9 illustrates the emerging business models for post-consumer recycling in urban China facilitated by internet technology. It signifies that business model innovation is booming there in recent years. Chapter 9 found three types of emerging business models related to waste handling: (1) community-based programs, (2) reverse logistic systems, and (3) pure internet solutions to bridge transactions between consumers and recyclers. The chapter describes that five elements are key to the viability of the business models in post-consumer recycling, including consumer convenience, producer traceability, recycler profitability, hybrid collection practices and information reliability. Chapter 9 finds that the introduction of the extended producer responsibility in China (originally from the European Union) has been the trigger for this wild bursting growth of business models. The chapter draws lessons from how other countries can foster business opportunities by adopting and translating policies into their waste management practices.

Chapter 10 stipulates that for both the urban government and corporate actors, informal recycling is a pivotal sector for waste management practices. The informal recycling sector helps these actors to shape value chains around the waste management and help achieve circular lifetime extension strategies for consumer goods. In this effort, the chapter reveals how the formal state-based policies in fact aim at the orchestration of the same waste management practices as the informal recycler sector caters for. Through this it demonstrates how the achievement of circularity faces extra challenges caused by this duality. The chapter describes how both sides are organized from different continuation logics. In effect, it is discussed why the informal recycling sector actually contributes more to inclusive and circular waste management practices than the state-orchestrated efforts. It unveils that inclusion also needs organic attention from ongoing practices to inform and shape policy.

Final and concluding Chapter 11 takes stock of the main theoretical and empirical take-aways from the book. It looks back at the various chapters, draws the key lessons from them and connects all the dots by offering a Plan-Do-Check-Act approach to bring the formal and informal sectors operating in the Urban Waste Management System (UWMS) together and examining what role they play in it, to which physical parts of the system they are linked and what policy changes and other inducements it would take to strengthen their productive participation within it. All too often, official policies ignore their essential contribution to both inclusion and circularity. More generally, it emphasizes the productive use of applying a systems approach for transitioning to an inclusive circular UWMS.

References

1. Coletto D, Bisschop L (2017) Waste pickers in the informal economy of the Global South: included or excluded? Int J Sociol Soc Policy 37(5/6):280–294. https://doi.org/10.1108/IJSSP-01-2016-0006
2. Egidi GL, Salvati S, Vinci (2020) The long way to tipperary: city size and worldwide urban population trends, 1950–2030. Sustain Cities Soc 60:102148. https://doi.org/10.1016/j.scs.2020.102148
3. Gutberlet J (2023) Global plastic pollution and informal waste pickers. Camb Prism Plast 1:e9. https://doi.org/10.1017/plc.2023.10
4. Hoornweg D, Bhada-Tata P, Kennedy C (2013) Environment: waste production must peak this century. Nature 502:615–617. https://doi.org/10.1038/502615a
5. Liu Z, Schraven D, De Jong M, Hertogh M (2023) Unlocking system transitions for municipal solid waste infrastructure: A model for mapping interdependencies in a local context. Resour Conserv Recycl 198:107180. https://doi.org/10.1016/j.resconrec.2023.107180
6. Mahardika S, Pratamo NBA (2024) The problem of plastic waste, a comparison of the systems in Indonesia and the Netherlands: The Urgency of state intervention and legal instruments. Environ Pollut J 4(1):934–949. https://doi.org/10.58954/epj.v4i1.172
7. Porras Bulla J, Rendon M, Espluga Trenc J (2021) Policing the stigma in our waste: what we know about informal waste pickers in the global north. Local Environ 26(10):1299–1312. https://doi.org/10.1080/13549839.2021.1974368
8. Rootes C (2009) Contesting toxics: struggles against hazardous waste. Environ Polit 18(2):287–291. https://doi.org/10.1080/09644010802682668
9. Schneider AF, Zeng X (2022) Investigations into the transition toward an established e-waste management system in China: Empirical evidence from Guangdong and Shaanxi. Curr Res Environ Sustain 4:100195. https://doi.org/10.1016/j.crsust.2022.100195

Open Access This chapter is licensed under the terms of the Creative Commons Attribution 4.0 International License (http://creativecommons.org/licenses/by/4.0/), which permits use, sharing, adaptation, distribution and reproduction in any medium or format, as long as you give appropriate credit to the original author(s) and the source, provide a link to the Creative Commons license and indicate if changes were made.

The images or other third party material in this chapter are included in the chapter's Creative Commons license, unless indicated otherwise in a credit line to the material. If material is not included in the chapter's Creative Commons license and your intended use is not permitted by statutory regulation or exceeds the permitted use, you will need to obtain permission directly from the copyright holder.

Exploring the Circular City: A Bibliometric and Definition Analysis

Quirien Reijtenbagh, Zhaowen Liu, and Daan Schraven

Abstract Having a high demand for materials and vast emissions, cities are ideal laboratories for exploring the circular economy. The circular city as a city label has attracted a lot of attention from academics and practitioners. However, it remains unclear what a circular city is in the context of the circular economy in the urban environment. To improve our understanding of how circular cities have been defined and developed by scholars, this book chapter reviews academic literature. First, we set a search scope of relevant literature using a bibliometric approach to build a database with 109 peer-reviewed articles. Then, we compile a final list of 28 definitions which are analyzed to find the dimensions of a circular city. These circular city dimensions are: system structure, components, principles, goals, and organizational characteristics. In sum, this chapter provides a comprehensive understanding of the current central themes of circular cities and their shortcomings.

Keywords Circular city · Circular economy · Sustainable development

1 Introduction

Cities are increasingly attracting residents because of their modern businesses and services, diverse employment opportunities, and accessible infrastructure. In 2020, 55% of the world's population were living in cities, and this proportion is expected to increase to 68% by 2050 [7]. Global-scale urbanization has witnessed economic

---

Q. Reijtenbagh (✉) · D. Schraven
Faculty of Architecture and the Built Environment, Delft University of Technology, Delft, The Netherlands
e-mail: Q.A.M.Reijtenbagh-1@tudelft.nl

D. Schraven
e-mail: d.f.j.schraven@tudelft.nl

Z. Liu
Faculty of Civil Engineering and Geosciences, Delft University of Technology, Delft, The Netherlands
e-mail: z.liu-8@tudelft.nl

© The Author(s) 2026
M. de Jong et al. (eds.), *The Inclusive Circular Economy*, Urban Sustainability,
https://doi.org/10.1007/978-981-96-6867-0_2

growth, but it has also inevitably led to massive resource consumption and production in cities [31]. Meanwhile, the convenience and diversity of urban lifestyles resulted to an increase in the amount and complexity of municipal waste generation [13, 24].

An alternative to this take-make-dispose economy is the circular economy, which proposes development strategies that extend the life of products and transform waste into resources [8, 14]. In practice, the circular economy aims to drive sustainable development by reducing the demand for (virgin) raw materials and by extending product use through innovating business models which involve and/or address a plurality of societal actors while minimizing environmental impact [16]. Besides, its main principles aim to dematerialize and diversify material supply, to ultimately relieve future resource scarcity issues [9].

Many countries and regions have adopted the circular economy as one of the key strategies for sustainable development [16]. As an economic system with high consumption patterns and emissions, cities are naturally the best laboratories for exploring the circular economy. For example, as a concept, the circular economy has already trickled down from strategic documents into the international level (e.g., EU Circular Economy Action Plan) and national level (e.g., national program of the Circular Economy), to the local level (e.g., city of Amsterdam). In scholarly literature, the use of the "circular city" as a city-label purporting "circular economy" in the urban setting, has been published a notable number of times, especially around 2020. However, a clear definition is yet absent [20], which may be partly due to a variety of definitions about the circular city found in the academic literature.

To address this unclarity, in this chapter we offer an overview of what the circular city is about. We do so by answering the question: *what is a circular city in the context of the circular economy in the urban environment?* First, by means of bibliometric analysis we identify various thematic elements, popular topics discussed amongst scientists, and points of convergence to demarcate how the circular city has been used in the academic literature. Second, we identify the compositional dimensions of a circular city by means of reviewing and recognizing patterns among existing definitions for 'circular city' (e.g., the dimensions or levels of the circular city).

The results show that the most important themes of the circular city include: adaptive reuse, indicators, nature-based solutions, and smart digitalization. In terms of dimensions of the circular city, these are: system characteristics (i.e., the traits of the circular city as a system), components (i.e., the entities or elements of a circular city), principles (i.e., the fundamental rules that guide action or decision-making toward a circular city), goals (i.e., for what reason or purpose is the circular city pursued), and the organization of a circular city (i.e., the way to establish the structure, processes, and relationships necessary for realizing a circular city). Based on these findings we formulated a conceptual framework of the circular city concept (Fig. 4).

2 Method

In this section, we describe the methodological steps for conducting the study. Firstly, we describe how we collected the relevant literature for the study. Secondly, we elaborate on the use of bibliometric techniques. Thirdly, we detail the steps of the content analysis on definitions, supported by an inductive coding process.

2.1 Literature Collection

For the literature analysis the Scopus database is used. Scopus indexes more journals than Web of Science, including more international and open access journals [1]. It is therefore able to provide literature on a wide variety of angles and disciplines. This is especially useful for the explorative nature of this chapter, where openness is warranted to find the diversity of dimensions that the concept of circular city may have received. In following the main question for this chapter, the query we deployed in Scopus reads:

TITLE-ABS-KEY ({circular city} OR {circular cities} AND {circular economy})

The query was conducted on March 4th, 2023, and was limited to (1) the last completed publication year, i.e. 2022; and (2) the most common academic language, i.e. English. With these limitations, we retrieved 109 articles, spanning between 2018 and 2022. For the thematic analysis, these articles were subject to bibliometric enquiry.

For the definitions analysis, we required that an additional criterion needs to be met for articles. It is argued that at the basis a proper definition needs to have a mature and scrutinized academic scholarly work behind it, i.e., journal articles or review articles. We ended up with a refined dataset of 84 articles.

We then searched definitions of the circular city in the full text of these 84 articles following four steps: (1) search for "def", "circular city" and "circular cities" in each publication; (2) read all the searching results; (3) extract the definitions; (4) remove duplicate definitions resulting from original citations. This resulted in a list of 28 original definitions (see Appendix). These identified definitions were subjected to a definitions analysis by coding in ATLAS.ti to identify the aspects that have been discussed in circular city definitions and to synthesize a conceptual framework of the circular city.

2.2 Data Analysis Approach

In this section we describe the analytical steps of the study. First, we describe the bibliometric analysis on the broader set of 109 articles retrieved. This is done to explore the associated themes to the circular city in the academic debate. Next, we explain the steps taken to analyze the 28 definitions retrieved from the set of 84 articles.

2.2.1 Bibliometric Analysis

Bibliometric literature review enables rapid analysis of keyword, author, and bibliographic information in large samples [22]. We applied bibliometric analysis in this study to quantitatively identify the development dynamics and leading trends of studies about the circular city. Specifically, keywords co-occurrence analysis and bibliographic coupling analysis can give insights on the conceptual evolution of a research field [25]. First, a keyword occurrence analysis was conducted to identify the ranking of associated terms to the circular city. Based on this ranking, a cut-off point was established to determine the high-frequency keywords. This is often used to discern the influential themes [18].

We pursue this enquiry with descriptive statistics, keyword-co-occurrence analysis, and cluster analysis. Regarding the descriptive statistics of the average publication year for these authors terms, which can show the degree of forward leaning of these concepts. For example, if the frequency of author keywords is counted per publication per year, then the average publication year for a keyword is the statistical middle of these counts expressed in years. It is counted as the average of the sum product of the count per year with the publication years, divided by the frequency of publications. For example, if an author keyword is published in three articles in the years 2020, 2022 and 2022, then the average publication year would be (1 × 2020 + 2 × 2022) / 3 = 2021,33. These results are shown in Fig. 1. Additionally, we wanted to understand whether these articles were used in a specific moment in time or continually throughout the observed period. For this we created boxplots using the median and first and third quartiles of frequency of occurrence of publication years per high-frequency keyword. The result is visually presented in Fig. 2.

With regards to the co-occurrence analysis, we counted the times that each possible pair of the high frequency keywords occurred together. This helped to produce a square matrix of all the co-occurrences. This was then used to create a social network graph in Fig. 3a.

With regards to the cluster analysis we used the same data set with the statistical software tool called SPSS. A cluster analysis helps to group keywords together based on their co-occurrences. The cluster analysis differs from the social network analysis by looking at the statistical correlations of the raw co-occurrences between keywords. The output of a cluster analysis is presented as a dendrogram, i.e., a tree-like representation of the found clusters (in Fig. 3b).

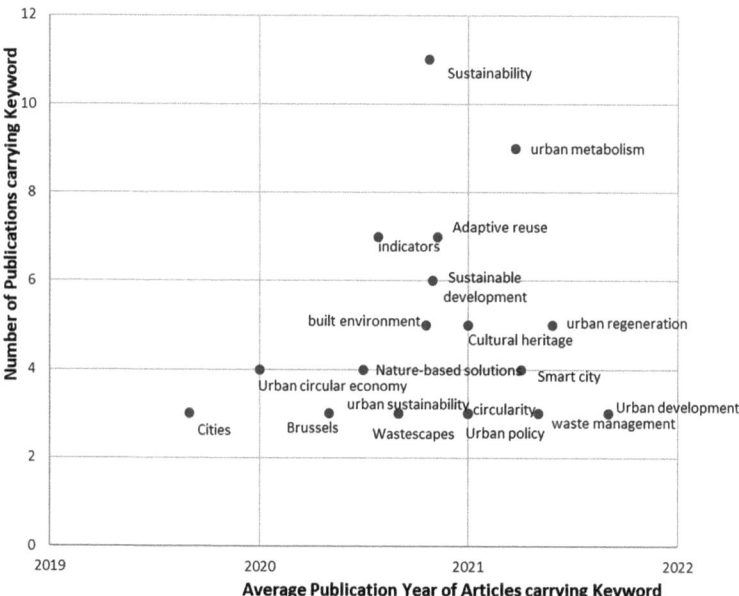

Fig. 1 Average Publication year versus number of publications per keyword

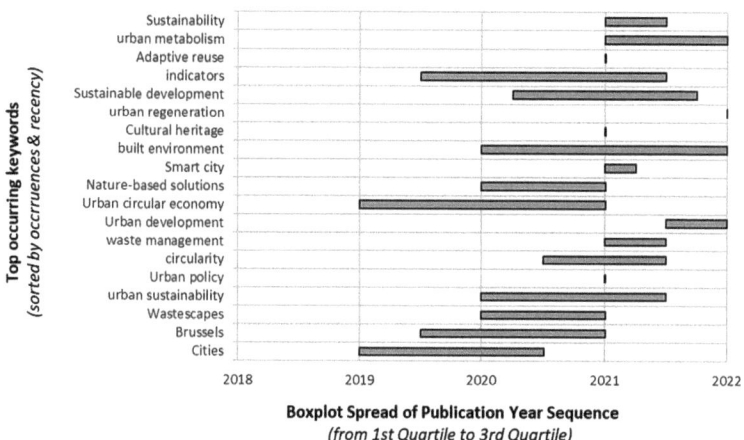

Fig. 2 Boxplot spread of Publication year sequence per keyword

2.2.2 Content Analysis of Definitions

We applied a qualitative coding approach to identify the elements of definitions. To ensure a proper interpretation of the results, two of the three authors with preconceived knowledge on the topic had carried out the coding process together. The purpose of our data coding procedure was to achieve the final goal: a coding frame that clearly represents all aspects of the circular city definition.

We based our coding process on the coding process for qualitative research recommended by Creswell [6, p. 244]. This procedure starts by an initial reading of the data, followed by segmenting and labelling the text, grouping codes and lastly, deriving themes. This coding procedure to find key elements (categories) was conducted in *ATLAS.ti*. The coding steps we conducted are presented in Table 1. We then used the output of the coding analysis (Table 4) to reexamine all 28 definitions to gather information on the type, amount, and content of (sub)categories included.

3 Results

In this section, we present and discuss the results from the bibliometric study. Then, we show and elaborate on the definitions and conclude with the conceptual framework.

3.1 Results from Bibliometric Analysis

In academic articles, authors often think of keywords that make their study easier to find. Therefore these keywords act as important descriptors of the essence of a field of study. Taking stock of this for circular cities, we conducted a keyword occurrence analysis. For this, one counts the number of single occurrences among the pool of articles. In this case, the total number of articles are 109 until the end of 2022. In order to find the associated terms for circular cities, we removed the three keywords that were actual search terms in our query, i.e. "circular city", "circular cities" and "circular economy". We then sorted the remaining terms from high to low.

Table 2 presents the top 19 ranking keywords with cut-off point for 3 occurrences. It shows that sustainability is the most associated term to circular cities, which is no surprise, provided the wide range of the scope of this term. More specific terms include: "urban metabolism" (9 hits), "adaptive reuse" (7 hits), "urban regeneration" (5 hits), "nature-based solutions" (4 hits), and "waste management" (3 hits) to describe certain strategies in the circular economy in an urban context. Terms like "built environment" (5 hits), "cultural heritage" (5 hits), "smart city" (4 hits), "urban development" (3 hits) and "urban policy" (3 hits) refer to the spatial nature of the city for such strategies (Fig. 3).

The keywords are occurring with a certain distribution over the years. The time-aspect is particularly useful to gain a better understanding of the recency and spread of used keywords. It can tell us more about the possible shifts in attention and association of circular city in studies as time progressed. We therefore created a scatter plot, by showing the total occurrence and its average publication year of articles that carry a certain keyword resulting in Fig. 1.

Table 1 Coding procedure of circular city definitions

Coding steps	Explanation	Outcomes
Step 1. Data preparation and familiarization	We prepare a clean dataset of definitions and preliminary scan reading of all the definitions	28 definitions of the circular city (Appendix)
Step 2. Set the coding rules (*collectively*) and conduct coding (*individually*)	Following Creswell [6], we apply the coding rules as: 1. every part of the definition needs to be coded, unless the parts of the definitions were very vague or unspecified in meaning (e.g., "generate accomplishment", "enhance the flexibility of the city and its people" and "complex system" and "social consumption") 2. if codes were overlapping (i.e., meaning the same) they were merged (e.g., "economic growth" and "economic prosperity" are merged into "economic prosperity")	120 codes from author A and 67 codes from author B
Step 3. Group codes (*individually*)	Codes with similar characteristics are grouped (e.g., actors mentioned in the definitions are categorized in the sub-category "stakeholders")	20 groups from author A and 16 groups from author B
Step 4. Merge groups into preliminary categories (*individually*)	Each author classifies the groups and identifies four categories. Each category may have two to three sub-categories	Categories from author A: system elements, system boundaries, goals, and principles; categories from author B: components, goals, circular city system, organization/means
Step 5. Finalize the categories (*collectively*)	This step involves the comparison of codes and preliminary categories from step 4, and the overall similarity is high. Some changes are made to create the final coding framework: 1. non-similar codes are added (18 codes were added), 2. non-similar categories and sub-categories are discussed and merged	Five aspects of the circular city that emerged from the coding process (Table 4)

Table 2 Occurrence analysis of most associated keywords to circular city

Rank & Keyword		# of occurrences
1	Sustainability	11
2	Urban metabolism	9
3	Indicators	7
4	Adaptive reuse	7
5	Sustainable development	6
6	Built environment	5
7	Urban regeneration	5
8	Cultural heritage	5
9	Smart city	4
10	Nature-based solutions	4
11	Urban circular economy	4
12	Urban development	3
13	Urban sustainability	3
14	Brussels	3
15	Waste management	3
16	Circularity	3
17	Wastescapes	3
18	Urban policy	3
19	Cities	3

Figure 1 shows that around 2020 frequently occurring keywords were "cities", "urban circular economy" and "Brussels". In 2021, the focus shifted more towards operationalization efforts, by studying "indicators" and "circularity" approaches associated with solutions or strategies deployable in a circular city such as "nature-based solutions", "adaptive reuse" and "urban policy". After 2022, terms like "urban development", "urban metabolism", and "urban regeneration" appeared in tandem with the concept of the circular city. This indicates that the term circular city was consolidated with older debates on transformations in the urban space.

Given that the average number of publications per year does not yet improve our understanding on the commonplace of the coined keywords, we further clarified this by visualizing the spread of the publication sequence of the various terms. We created boxplots of the publication sequence per keyword, ranging from the 1st to the 3rd quartile of the sequence. Usually boxplots are known for their whiskers at the sides of these quartiles, but provided the short span of publication period (5 years) we removed these from the figure. This helped in focusing the message of the visual on the identification of the continued and short-lived momenta of the keywords in the circular city literature. In this way Fig. 2 reveals that some terms have been packed while others were more regularly spread across the years. To compare positioned keywords easier, we sorted the keywords across their reading direction in Fig. 1, first by their occurrence and then by their recency.

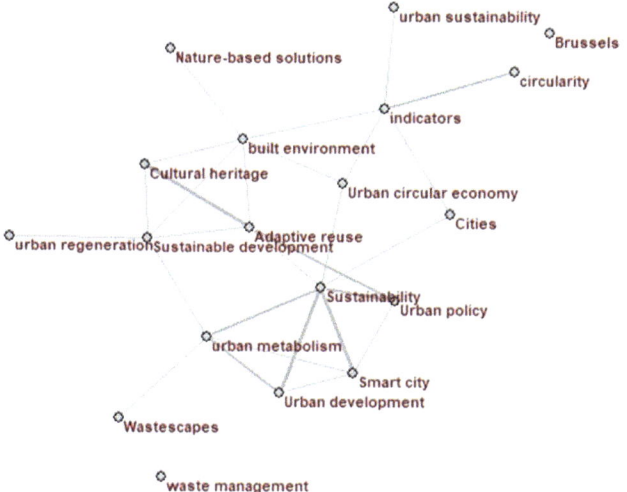

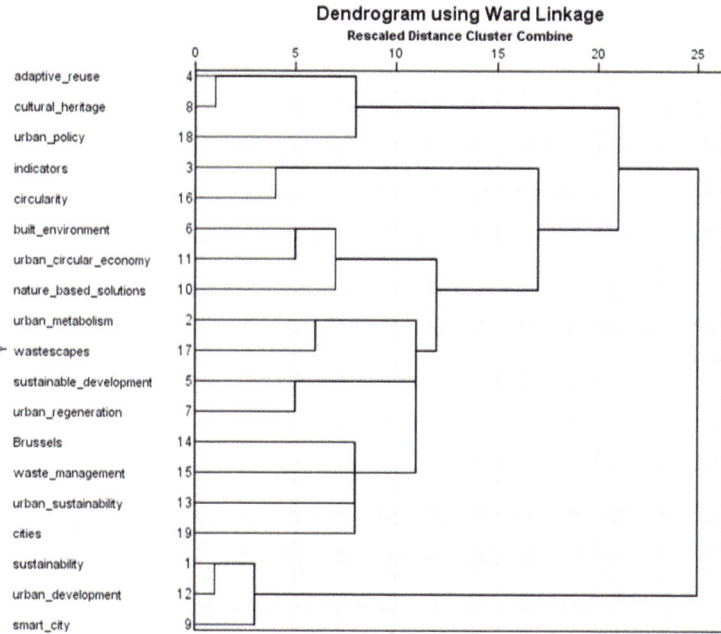

Fig. 3 a. Social network analysis. **b.** Cluster analysis

The largest spread over the past five years occurred for the terms "indicators", "built environment" and "urban circular economy" (Fig. 2) which indicates their longevity in the academic debate. Interestingly, the terms related to solutions appear to have been published quite close after one another, as revealed by the narrow spread of "adaptive reuse", "cultural heritage", "smart city" and "urban policy" but no conclusions can be drawn about the underlying reasons.

In the previous descriptive statistics, we focused mainly on the keywords that were mentioned in the studied literature most frequently. By just relying on these terms, we would omit other terms that studies have associated to the circular city label. To study the extent that the label has been converging with the most frequent occurring words, we separated the keywords that contributed to the divergence (i.e., the sum of all single occurrences) from the convergence (i.e., the sum of all > 1 occurrences) of the circular city label. These were then separated across the publications years to elucidate the change in the extent of divergence versus convergence (Table 3).

The converging ratio increased from 22% in 2018 to 41% and 30% in 2021 and 2022, respectively, indicating that associations to the circular city concept had a steady pace towards a consolidated understanding.

Next, we place our focus on high frequency keywords to represent the core content of the studied literature, to offer an insight into the potential directions that the debates are moving toward, and to illuminate the core aspects of the circular city. The number of high-frequency keywords can be identified by following the proposed model by Donohue (1973). In this model, the tail of sorted single keywords help to determine the statistically expected cut-off point for the high frequency keywords. The formula for calculating this is:

$$I_n = \frac{1}{2}\left(-1 + \sqrt{1 + 8 \times I_1}\right)$$

In the formula I_1 represents the number of keywords that occurred just once and I_n represents the total number of high-frequency keywords. A total number of 304 articles occurrence only once. That meant that for this dataset of the circular city literature this equals a maximum of 24 keywords. Table 2 shows that the position 12 through to 19, have 3 hits. The 2 hits exceed the 24 keywords limit, and hence we set the cut-off at the top 19 keywords for further analysis of the underlying relational structure among them.

Figure 3a shows the results of the social network analysis conducted on frequently occurring keywords surrounding the concept of circular city. A few groups can be distinguished. The first group that can be identified, concerns urban aspects like "urban metabolism", "urban policy", "urban development" and "smart city". These are all tied strongly tied to the keyword "sustainability". The second group surrounds "cultural heritage" and "adaptive reuse". These focus strongly on the changes in

Table 3 Percentage diverging/converging keywords

Years	2018	2019	2020	2021	2022
Diverging keywords	14	44	53	91	102
Converging keywords	4	17	17	63	44
Converging ratio	22%	28%	24%	41%	30%

Table 4 Results on aspects of the circular city that emerged from the coding process

Categories	Sub-categories	Description	Code examples
Circular city system description	Characteristics	Terms that characterize the circular city. This could be adjectives but also nouns that describe an ability of the system	Adaptive; regeneration; innovation; technical system; socio-ecological system
	System boundaries	Terms that describe the boundaries of the circular city. These are directly mentioned boundaries, but can also be related nouns	Functional boundaries, nature boundaries, spatial boundaries, city-region; bio-geo-physical unit
Circular city components	Stocks	Stocks are the resources that are accumulated and stored within the city, including man-made, social, cultural, human, natural, and digital capital	Built environment; mobility; industry; infrastructure; cultural heritage capital; digital technology
	Flows	Flows are the movement of these resources through the city's economy	Resource flow; waste flow; information flow; energy flow
Circular city principles	10R+ strategies	Terms about 10R frameworks, including, as well as other Rs that have been mentioned	Recover; reduce; redesign; repair; reject; re-naturalization
	Other circular economy principles	Terms about circular economy principles other than R strategies, including prolonging the lifespan of products, waste management, and renewable-energy	Waste elimination; waste to resource; renewable energy; closing, narrowing, slowing the loops
Circular city goals	Economic development	Terms that related to the ideas on economic development in the circular city	Economic prosperity; value creation; value extension; efficiency improvements
	Environmental quality	Terms that related to the use of natural resources in the circular city	Ecological impact; environmental benefits; renewable natural resources
	Social well-being	Terms that related to social well-being in the circular city. This could both be adjectives	Healthy; human centered; inclusive; social-wellbeing; increase habitability

(continued)

Table 4 (continued)

Categories	Sub-categories	Description	Code examples
Circular city organization	Stakeholder engagement	Terms that related to organizing stakeholders in the design and implementation of circular city	Collaboration; stakeholders; institutions; knowledge sharing; integrated network
	Circular industry development	Terms that related to organizing businesses that contributes to a circular city	Industrial symbiosis; business model; logistical services
	Circular actions, practices and initiatives	Terms that related to practices and behaviors in realizing a circular city	Circular economy initiatives; circular economy practices; behavior change

the built environment, and then more clearly, on the management of historical or existing buildings. Finally, there is a recognizable group on "indicators" and "circularity", which clearly share the emphasis on clarifying how to measure the circular performance of the city.

Figure 3b shows the cluster analysis as a dendrogram. There exist many similarities between the formed clusters from this figure confirming three distinct emphases in the debate. First, the keywords "sustainability", "urban development", and "smart city" contain the dominant terms. Also "adaptive reuse" and "cultural heritage" form a strong bond regarding how to deal with the existing building stock. Third, the keywords "indicator" and "circularity" form a separate topic on measuring performance. The keywords which capture various urban aspects seem to be interrelated but it is striking to note a detachment of waste-related topics like waste management and wastescapes from the circular city concept.

3.2 Results from Definitions Analysis

3.2.1 Finalized Coding Framework

In this section we present the result of content analysis of definitions. Following the coding procedure that introduced in Sect. 2.2.2, five main categories were found in the definitions, including: (1) *circular city system description,* (2) *circular city components,* (3) *circular city principles,* (4) *circular city goals, and* (5) *circular city organization.* Table 4 displays the five aspects, their subcategories with descriptions, as well as the examples of codes in each sub-category.

Table 5 Occurrence of categories and sub-categories in definitions

Definitions that mention X of total	Frequency in definitions	Percentage in definitions (%)
Circular city definitions (total)		
-	28	100
Circular city as a system		
System characteristics	8	29
System boundaries	8	29
Circular city components		
Circular city stocks	18	64
Circular city flows	7	25
Circular city principles		
R-strategies	15	54
Other CE principles (e.g., close, narrow, slow the loops)	7	25
No principles	6	21
Circular city goals		
No goals	18	64
At least one goal	10	36
Include three dimensions of goals (ED, EQ, SWB)	5	18
Economic development (ED)	5	18
Environmental quality (EQ)	10	36
Mention social well-being (SWB)	6	21
Circular City organization		
Stakeholder engagement	12	43
Circular industries	7	25
CE actions/practices/initiatives	9	32

3.2.2 Analysis of Five Aspects of Circular City

Because of the length of final list of definitions, the list is found in the Appendix 28 in the end of this chapter. For proper referencing and increasing text readability, the numbers in the first column of Appendix correspond to the numbers in brackets in the text. For example, reference [1] in the text refers to [1] in the Appendix, that is the definition provided by Vanhuyse et al. [29] provided at page 3 of their article (Table 5).

Circular City as a System

Eight definitions (29%) refer to different type and reach of boundaries of the circular city. For example, definition [7] indicates that the circular city has *spatial and functional boundaries*. Some definitions give a notion of a cross-boundary nature of the circular city, due to input and output material flows beyond the city borders. These definitions indicate that the circular city is a *part of regional development* [2], or a *city-region* [12] implying a *multi-scalar* relation to creating circular material flows, as a result of collaboration between actors from the *hinterlands* [19].

Eight definitions characterize the circular city as a 'system'. These notions range from *urban (eco)system* [9,11] to *self-sustainable urban system* [12, 19] to *sociocultural system* [21] and *socio-ecological system* [7]. Often the circular city is described as a *regenerative urban system* [2, 7, 12, 20]. The characteristics of the circular city differ between the definitions. For example, the circular city is regarded as a *movement* [14], is *smart* [13], *self-sustainable* [12, 19] and regenerative [2, 7, 12, 18, 20] and is a *place for innovation to trial and scale-up solutions* [9].

Circular City Components

From the 20 definitions, around 61% mention resources that are accumulated and stored within the city. These definitions pose that the circular city is about a different (circular) use of resources either in the form of regenerating capital or in the form of creating circular resource flows (and efficient energy flows).

Stocks (i.e., cumulated resources) are referred to as human, man-made, social, and natural capital [16, 17] or to social, environmental, and economic resources [18]. Some definitions include specific territorial resources in their definition, like *infrastructure* [9, 15, 21, 22], *housing* [9] and *land* [21] that can be *redesigned, reused or recycled*. Other key elements identified in the definitions are *clean and efficient mobility* [10, 21] and *(digital) technology* to empower circular solutions [13, 20]. Human capital is mostly mentioned in the terms of certain stakeholders, i.e., *social* [7, 23], *economic actors* and *institutions* [13, 27], as well as *industrial actors* [6]. Resource flows are mostly referred to as *material, waste, and energy flows* [1,10]. One definition even included specifics on material flows, mentioning *food, nutrition,* and *water* [9].

Circular City Principles

About 54% of the definitions mention R-strategies. Besides the 10R-strategies of Potting et al. [23, p. 5] (i.e., *refuse, rethink, reduce, reuse, repair, refurbish, remanufacture, repurpose, recycle, and recover*). Three additional strategies are also highlighted in some of the definitions: *redesign of industry, infrastructure and logistics* [21], *re-naturalizing the city* [17], *and regeneration* [2, 7, 11, 12, 16, 17, 18, 20, 27]. Regeneration is mentioned in three ways: first, as a characteristic of the economy of the circularity city, being a *regenerative economy* that sustains the ability of natural ecosystems [2]. Second, as a characteristic of the whole urban system [7, 11, 12]. Third, as the regeneration of natural [16, 17], cultural, social, economic, produced, and human *resources or capital* [17]. Other R-strategies mentioned were *redesign* and *re-naturalize*. Redesign refers to the *"redesign of infrastructure, logistics and industry"* [21]. Re-naturalize refers to creating more nature in cities and "sharing" resources [4, 9].

About 25% of the definitions include other circular economy principles as overarching strategies for the application of *circular economy principles* in the circular city [20] or more specifically for *"closing, narrowing or slowing the loops"* [4, 6, 8, 10, 13]. Two definitions explicitly note the idea behind the application of circular economy principles as *"using resources as long as possible at their highest value"* [9, 18].

Circular City Goals

From all definitions, 64% of the definitions do not include goals of the circular city. Of those which do, the goals mentioned are related to the three dimensions of sustainable development: economic prosperity, social well-being, and environmental quality.

The most often mentioned goal of the circular city is environmental quality (e.g., restoration, regeneration, conservation). Examples include the use of local renewable natural resources [4], the preservation of the ability of natural ecosystems to secure production and maintain diversity [2], the regeneration of natural capital [16, 17], and the reduction of the environmental footprint [3].

With respect to economic development as a goal, half of the definitions include a specific idea on economic development within the circular city. These are: *having economic prosperity* by a *regenerative economic system* [2], *having economic growth* by reducing the environmental footprint [3], *being competitive* by using resources as long as possible in their original form [9], *providing economic benefits* by being a circular city [14] and lastly, *creating positive economic externalities* by using regenerative capacity of the system [18].

Lastly, 21% definitions include the goal of social well-being for creating a certain *social well-being* [2, 27], *ensuring well-being for locals* [3], and striving for an *inclusive and healthy city* [9].

Circular City Organization

Many definitions refer to various ways of organizing circular cities by engaging stakeholders, promoting circular industries, and conduct circular actions, practices, and initiatives. 43% definitions mentioned *stakeholder engagement*. For example, this included mentioning *stakeholder cooperation* (e.g., including citizens, community, business and knowledge stakeholders) [8, 9, 18, 19, 23] and *knowledge sharing*

[9, 15]. With respect to circular industry development, definitions refer to *(industrial) symbioses and clustering* [18, 19], the application of *circular business models* [6, 11, 27, 28], and business activities related to organizing recycling, remanufacturing, storage, and *(reverse) logistics* [26]. Some definitions also show that creating a circular city involves actions on "*scaling up integrated networks*, retrofitting existing businesses, and creating new operational practices across scales" [27], and drawing *lessons from best practices* [24].

Lastly, the circular city is established by *governance on multiple levels* [21]. This refers to the cross-boundary nature [27] of material flows and waste and material regulations, that require *multi-level policy integration* [21]. Hence, local governments have limited capability to adjust regulations when these are governed at the regional, national, or international level.

Based on the findings from definitions analysis, we synthesize a conceptual framework of the circular city (Fig. 4).

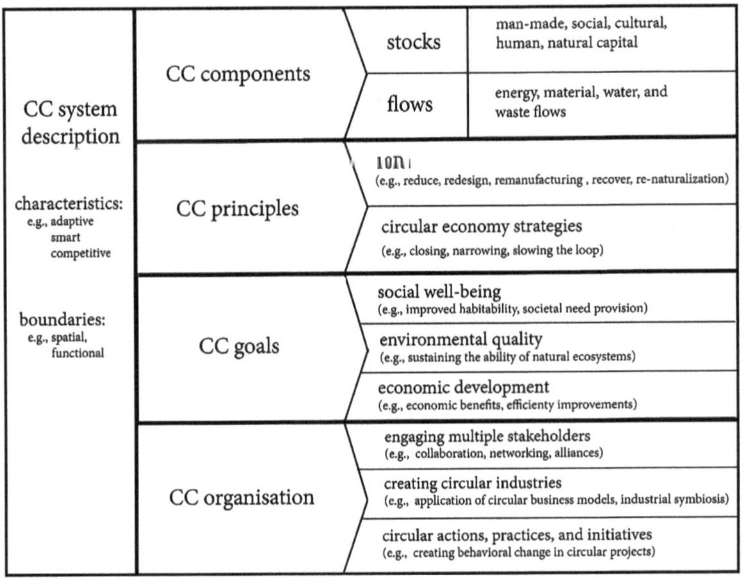

CC system description	CC components	stocks	man-made, social, cultural, human, natural capital
		flows	energy, material, water, and waste flows
characteristics: e.g., adaptive smart competitive	CC principles	10R's (e.g., reduce, redesign, remanufacturing, recover, re-naturalization)	
		circular economy strategies (e.g., closing, narrowing, slowing the loop)	
boundaries: e.g., spatial, functional	CC goals	social well-being (e.g., improved habitability, societal need provision)	
		environmental quality (e.g., sustaining the ability of natural ecosystems)	
		economic development (e.g., economic benefits, efficiency improvements)	
	CC organisation	engaging multiple stakeholders (e.g., collaboration, networking, alliances)	
		creating circular industries (e.g., application of circular business models, industrial symbiosis)	
		circular actions, practices, and initiatives (e.g., creating behavioral change in circular projects)	

Fig. 4 A conceptual framework of the circular city

4 Discussion

4.1 Discussion on the Bibliometric Results

The bibliometric results highlight a few notable things. First, a few conjoined and clearly identifiable themes appear to have received attention. The measurement of circularity with indicators is a very clear theme that was called out through the keywords. This debate has been ongoing for at least the past two years, making it a fundamental aspect of the conceptual development of the circular city.

Also, the concept of adaptive reuse and cultural heritage appear to be anchored in the circular city domain. This speaks to a clear strategy for keeping buildings functioning, if not for one specific role, then for another. The idea that buildings can be quickly adapted for a plethora of functions is therefore a viable route. These topics were published in a very narrow window around 2021 which could be an indication of very specific and time-bound research.

Interestingly, waste management has been mentioned only fairly recently, but it has not been manifested as a strong link with other circular city author keywords. This signals that the subject of waste management is young in the debate on circular cities, but not yet firmly connected with other debates, like urban regeneration and urban development.

4.2 Reflections on Definitions of the Circular City

The coding analysis revealed five aspects of the circular city: the system, the components, the principles, the goals, and the organization. In this section we provide a critical review on these definitions.

4.2.1 Regional Boundaries Contrast with Cross-Region Flows

Many definitions describe the circular city as a system with spatial/regional/geographical boundaries. Campbell-Johnston et al. [5] concluded that circular cities aim to create more local material flows of formerly dispersed flows. While this is contrary to the cross-city-boundary stocks and flows as components of a circular city, it does bring up the discussion of whether it is feasible and favorable to reach circularity within the city boundaries. Since some definitions also include cross-boundary aspects. For example, the recycling of metals and critical raw materials is an (inter)nationally organized practice that goes beyond the city boundaries [26].Thus, it can be concluded that circular city boundaries could contrast with the traditional notion of a city as a contained entity with defined boundaries due to the cross-boundary nature of material and energy flows.

Furthermore, we observe that the most recognized characteristic of a circular city is that of "regeneration", referring not only to natural capital, but also to cultural and social capital which are also present beyond the geographic city boundary.

Lastly, regulations that are enforced on multiple levels (i.e., regional or (inter)national) require multi-level governance and collaboration with stakeholders outside the city boundaries.

4.2.2 Principles Do Not Align with Multidimensional Goals

Almost all definitions that include circular city goals emphasize the importance of environmental quality. This is in line with the strong sustainability view that natural capital is essential for providing the well-being of the (future) population and cannot be replaced by productive capital [4]. Furthermore, it supports the view that a circular economy in cities is a means for transitioning towards a society that can achieve social well-being, environmental quality, and economic prosperity [30].

Most definitions are explicitly referring to the application of circular economy principles, such as looping strategies or 10R-strategies in the city. These principles prioritize resource efficiency in economic systems over other goals of circular city, such as social-wellbeing. To achieve social goals of circular cities such as inclusion, equity, and health, the guiding principles of the circular city must go beyond the limits of the circular economy to address potential trade-offs and unintended consequences.

What stands out from the results is that circular city is relatively often characterized as a regenerative system. Regeneration can refer to the regeneration of abandoned, degraded, or deserted places by creating new purposes or through redesign [12] or urban ecosystem restoration by creating green and blue infrastructures or natural capital in the city to provide ecosystem services (e.g., urban heat effect reduction) [31]. Interestingly, though regeneration was referred to in multiple ways, *"regenerating the loop"* was not recognized as a circular city principle. This is not in accordance with the latest insights in academia on the circular economy strategies, as these include economy: closing, narrowing, slowing, and regenerating the resource loops [3, 10, 15].

4.2.3 Biased Role Recognition of Stakeholders

Definitions of the circular city mention several urban stakeholders ranging from industrial actors to local inhabitants, as being part of the circular city. In general, actors in industries and businesses are described as key actors since they take a lead in the production process and business models [28, 29, 32]. Meanwhile local governments are politically empowered as circular policy makers and project leaders [5, 21].

However, the *citizen* as part of the circular society is described as a rather passive stakeholder. The circular city is considered a means to provide well-being for local inhabitants, rather than local inhabitants being active contributors to creating the circular city. This brings forth an implicit bias on different stakeholders whereby some are described as drivers of a circular transition and others are only passive actors or participants. It is not in accordance with the modern view on circular economy that citizens need to become active participants [19] and are expected to fulfill certain roles (e.g., social initiative organizers,green consumers) with circular behaviors[11, 17].

5 Conclusion

5.1 *Academic Contribution*

In this chapter, we addressed the question: *what is a circular city in the context of the circular economy in the urban environment?* With bibliometrics, we were able to show that a circular city is related to a few circular strategies for reducing the material use. The analysis suggests that the circular city is about keeping the built environment operating through adaptive reuse, by ensuring the adoption of nature-based solutions in the urban setting, while ensuring that waste management is addressed. A circular city also looks into the urban development practices and strives for urban regeneration.

From the definitions analysis, we uncovered that there are five dimensions of a circular city: the system description, the components, the principles, the goals, and the organization. These dimensions provide us with the understanding that the city materializes in the form of diverse forms of stocks and flows of materials, waste, water and energy. The way these flows and stocks are made circular is through circular principles or R-strategies. At the same time, a circular city pursues social wellbeing, environmental quality, and economic development. In order to achieve these goals, the city is organized by engaging with multiple stakeholders, create circular industries through circular business models and fosters circular actions, practices, and initiatives.

Furthermore, we reached to a few key conclusions on the circular city. First, the inclusion of "hard" city-scale boundaries into the definition contradicts the very idea of exchanging waste and material flows across the city boundaries. Furthermore, the principles of the circular city in terms of the strategies that it can deploy, do not necessarily contribute directly to all of the circular city goals. Such strategies can be reviewed to which extent these improve the city in all three sustainable development dimensions. Finally, the debate on engagement of certain actors or stakeholder, leans mainly towards industrial and governmental actors, rather than towards the citizens who are meant to be active participants in creating a circular economy. Hence, a perspective on city boundaries and citizen inclusion is still missing in the conceptualization of the circular city.

5.2 Practical Contribution

Based on our findings, we encourage practitioners concerned with the circular economy in the urban context to stimulate circular strategies that not only concern closing, narrowing, and slowing, but also regenerating the loop. Additionally, we think that there is much potential in mobilizing local capital and engage local/regional actors (government, business, and citizens) to sustain the circular economy. Also, striving towards creating a 'circular city' should rather be perceived as an instrument to create a sustainable society and environment, not as a goal in itself.

Lastly, based on the dimensions of the circular city, the following question can operate as a guidance for building a conceptual understanding of the circular city:

- For which flows would we like to execute the circular economy principles in our city?
- How is the implementation of circular economy principles organized and which stakeholders are involved?
- How can (local) capital (e.g. human, social, physical, natural etc.) help to execute these circular principles?
- What governance (e.g., regulations) is required and on which level?

5.3 Limitations

The limitations of this study are mainly reflected in the fact that the collection of definitions of the circular city focused only on the academic literature. However, the circular economy is not only an emerging field in academic literature but also in practice, the addition of definitions that are present in non-scholarly/grey literature would help to improve/validate the results. To compensate for this shortcoming, the follow-up study can include non-scholarly definitions, such as descriptions of circular cities from policy documents, to gather more insights in the aspects of the circular city.

Appendix 28 Definitions of the Circular City

No	References	Page	Definition
1	Vanhuyse, F., Fejzić, E., Ddiba, D., & Henrysson, M. (2021). The lack of social impact considerations in transitioning towards urban circular economies: a scoping review. Sustainable Cities and Society, 75, 103394	p. 3	Circular cities are cities that apply any of the 10R frameworks, within one or multiple industries, to close, slow and/or narrow the material and energy and waste flows within their geographical area
2	Nurdiana, J., Franco-Garcia, M. L., & Heldeweg, M. A. (2021). How shall we start? The importance of general indices for circular cities in Indonesia. Sustainability, 13(20), 11168	pp. 14, 15	A defined area within a regenerative economic system, with far-reaching regional strategic development that continuously sustains the ability of natural systems to remain productive and diverse, functions at a defined level of social well-being and that implies economic prosperity
3	Gravagnuolo, A., Girard, L. F., Kourtit, K., & Nijkamp, P. (2021). Adaptive re-use of urban cultural resources: Contours of circular city planning. City, Culture and Society, 26, 100416	p. 7	Circular cities strive to reduce their environmental footprint while ensuring economic growth and wellbeing for local populations
4	Paiho, S., Mäki, E., Wessberg, N., Paavola, M., Tuominen, P., Antikainen, M., ... & Jung, N. (2020). Towards circular cities—Conceptualizing core aspects. Sustainable Cities and Society, 59, 102143	pp. 6, 7	The circular city *is based on closing, slowing and narrowing the resource loops as far as possible after the potential for conservation, efficiency improvements, resource sharing, servitization and virtualization has been exhausted, with remaining needs for fresh material and energy being covered as far as possible based on local production using renewable natural resources*
5	Lakatos, E. S., Yong, G., Szilagyi, A., Clinci, D. S., Georgescu, L., Iticescu, C., & Cioca, L. I. (2021). Conceptualizing core aspects on circular economy in cities. Sustainability, 13(14), 7549	p. 13	A circular city is a city that functions through the usage of circular economy practices
6	Williams, J. (2021). Circular cities: what are the benefits of circular development?. Sustainability, 13(10), 5725	pp. 1, 2	Within this literature, circular cities are defined as those in which urban industrial actors adopt closed-loop production processes and business models
7	Williams, J. (2021). Circular cities: a revolution in urban sustainability. Routledge	pp. 14, 15	A circular city is a socio-ecological system, consisting of a bio-geo-physical unit and its associated social actors and institutions. It is a complex, regenerative and adaptive system, delimited by spatial and functional boundaries, surrounding an ecosystem

(continued)

(continued)

No	References	Page	Definition
8	Paiho, S., Wessberg, N., Pippuri-Mäkeläinen, J., Mäki, E., Sokka, L., Parviainen, T., … & Laurikko, J. (2021). Creating a Circular City–An analysis of potential transportation, energy and food solutions in a case district. Sustainable Cities and Society, 64, 102529	p. 187	A circular city constitutes: "one that practices CE principles to close resource loops in collaboration with its stakeholders to accomplish a future-proof city" (Prendeville et al., 2017, p. 187)
9	Definition from Holland Circular Hots pot (2019), available at: https://hollandcircularhotspot.nl/w-content/uploads/2019/04/HCH-Brochure-20190410-web_DEF.pdf (accessed 3 December 2019)	p. 6	A circular city is a resilient, healthy and competitive; able to provide for all the societal needs of its citizens within the natural boundaries of the Earth. Core elements of circularity are embedded within each key urban system; from water, to housing and infrastructure, to food and nutrition. Much like in a circular economy, in a circular city, resources are kept at their highest potential for as long as possible, through sharing, reusing, repairing, remanufacturing and recycling. Yet a city is inherently a human place; fostering collaboration and innovation to test and scale the solutions to create a truly inclusive, healthy and thriving place for all
10	Fusco Girard, L., & Nocca, F. (2019). Moving towards the circular economy/city model: which tools for operationalizing this model?. Sustainability, 11(22), 6253	p. 4	– a built environment designed in a modular and flexible way; – renewable-energy systems and efficient use of energy; – an accessible, economical, clean and effective urban mobility system; – recycling and transformation of waste into a resource; – production systems that encourage local loops closure and waste minimization
11	Williams, J. (2019). Circular cities. Urban Studies, 56(13), 2746–2762	p. 2759	A circular city is about a great deal more than creating a circular economy and circular business models within the urban context. It is about the regeneration and renewal of complex urban ecosystems
12	Gravagnuolo, A., Angrisano, M., & Fusco Girard, L. (2019). Circular economy strategies in eight historic port cities: Criteria and indicators towards a circular city assessment framework. Sustainability, 11(13), 3512	p. 17	The concept of a "circular city" or a "circular city-region" derives from the circular economy model applied in the spatial territorial dimension. It can be associated with the concept of a "self-sustainable" regenerative city. Later they mention: Circular cities can be seen as regenerative and self-sustainable systems

(continued)

(continued)

No	References	Page	Definition
13	Meskers, C., Caffarey, M., & Van Camp, M. (2019). Circular cities, E-mobility and the metals industry—a world in transition. In REWAS 2019: Manufacturing the Circular Materials Economy (pp. 313–318). Springer International Publishing	p. 313	Only in the abstract: Circular cities integrate all aspects of life, connecting across people, economic actors, institutions and geographies. Circular cities are powered by renewable energy and responsible, sustainable materials; have closed resource cycles and are smart. The technical, industrial, economic, cultural and social systems meet, interact and challenge each other
14	Krysovatyy, A., Zvarych, I., & Zvarych, R. (2018). Circular economy in the context of alterglobalization. Journal of International Studies Vol, 11(4)	p. 188	A circular city is a movement which promotes and uses a systemic way of thinking that can provide economic, social and environmental benefits to cities, while maintaining economic rationale
15	Cohen, J., & Gil, J. (2021). An entity-relationship model of the flow of waste and resources in city-regions: Improving knowledge management for the circular economy. Resources, Conservation & Recycling Advances, 12, 200058	p. 1, 2	The Circular Cities Hub1 defines a Circular City as a place where (1) "resources can be cycled between urban activities, within city regions" and (2) "cities can be designed so that land and infrastructure can be reused/recycled over time". One area of spatial planning that can directly contribute to (1) is waste management (Gravagnuolo, Angrisano, and Girard 2019; ESPON (European Spatial Planning Observation Network) 2019), and solving this problem requires a holistic approach that integrates knowledge from different domains
16	Bosone, M., & Ciampa, F. (2021). Human-Centred Indicators (HCI) to Regenerate Vulnerable Cultural Heritage and Landscape towards a Circular City: From the Bronx (NY) to Ercolano (IT). Sustainability, 13(10), 5505	p. 3	The circular and human-centred city is able to regenerate all forms of the existing cultural heritage capital (natural, manmade, cultural, social, economic and human) [8, 20] as key factors for achieving the sustainable development goals
17	Gravagnuolo, A., Micheletti, S., & Bosone, M. (2021). A participatory approach for "circular" adaptive reuse of cultural heritage. Building a heritage community in Salerno, Italy. Sustainability, 13(9), 4812	p. 26	The circular economy/circular city model implies also the re-naturalization of cities and the regeneration of the natural capital

(continued)

(continued)

No	References	Page	Definition
18	Bosone, M., De Toro, P., Fusco Girard, L., Gravagnuolo, A., & Iodice, S. (2021). Indicators for ex-post evaluation of cultural heritage adaptive reuse impacts in the perspective of the circular economy. Sustainability, 13(9), 4759	p. 13	Further study of Gravagnuolo et al. (2021) specified the conceptual evaluation framework placing CHAR in the perspective of the circular city model, identifying three main critical drivers or "building blocks" of circularity: – a "regenerative capacity" linked to the self-regeneration of the cultural assets, as well as of the economic, environmental and social resources needed for its maintenance over time (in analogy with the circular economy principle of extending the use value of resources in the largest time horizon possible) – a "generative capacity", linked to the net positive economic, environmental and social externalities generated in the area/territory—which in part come back to the heritage asset – a "symbiotic capacity", linked to the cooperation and collaboration approaches that enable a more efficient use of resources (such as those realized in "industrial symbioses"), as well as clustering processes in the territory (implementing an "economy of relationships")
19	Marin, J., Alaerts, L., & Van Acker, K. (2020). A materials bank for circular leuven: How to monitor 'messy' circular city transition projects. Sustainability, 12(24), 10351	p. 2	Acknowledging circular cities' multidimensionality related to the culture of cooperation, synergies, and symbioses that are key to the self-sustainability of urban and territorial systems complicates assessments that go beyond the 'materials and energy' dimensions in the circular urban economy Gravagnuolo et al. [12]. And later they mention: circular city projects are multiscalar, material objects or urban projects are interdependent with a multitude of hinterlands throughout the materials chain (authors, p.2)
20	Baganz, G., Proksch, G., Kloas, W., Lorleberg, W., Baganz, D., Staaks, G., & Lohrberg, F. (2020). Site Resource Inventories–a Missing Link in the <? xmltex\break?> Circular City's Information Flow. Advances in Geosciences, 54, 23–32	p. 24	The circular city (CC) generally applies the concept, principles and functions of CE and is thereby enabled by digital technology and foundationally designed as a regenerative and even restorative urban living system (Sukhdev et al., 2019)

(continued)

(continued)

No	References	Page	Definition
21	Campbell-Johnston, K., ten Cate, J., Elfering-Petrovic, M., & Gupta, J. (2019). City level circular transitions: Barriers and limits in Amsterdam, Utrecht and The Hague. Journal of cleaner production, 235, 1232–1239	p. 1233	A 'circular city', the newest iteration of urban sustainability initiatives, increases the 'added value' of urban metabolism by building on industrial ecology and integrating and redesigning infrastructure, logistical services, industries, and the socio-cultural system at multiple levels of governance, including more recently on social consumption
22	Ancapi, F. B., Van den Berghe, K., & van Bueren, E. (2022). The circular built environment toolbox: A systematic literature review of policy instruments. *Journal of Cleaner Production*, 133918	p. 2	In general, a circular city has the goal of improving the ecological impact of existing in- and out-going flows of materials and energy in urban buildings and infrastructures by making them as circular as possible
23	Paoli, F., Pirlone, F., & Spadaro, I. (2022). Indicators for the Circular City: A Review and a Proposal. Sustainability, 14(19), 11848	p. 1	The topic of the circular city is currently much debated in the literature and is seen as one of the possible solutions for achieving sustainability in urban areas. The transition to circular cities is at the center of this debate. Specifically, this transition involves the creation of an integrated city in which the principles of circular economy are applied to all local government divisions, in a process facilitated by political initiative and support that through good example promotes change among residents and various stakeholders. Thus, the basis of this vision is the employment of circular economy ideals, which include in their foundational design the concepts of second use, remanufacturing, efficient use of resources, elimination of waste, avoidance of toxic materials, and improving and making sustainable waste management through the utilization of the 9Rs strategy (reduce, reuse, recycle, recover, reject, repair, refurbish, remanufacture, and reuse)
24	Nocca, F., & Angrisano, M. (2022). The Multidimensional Evaluation of Cultural Heritage Regeneration Projects: A Proposal for Integrating Level (s) Tool—The Case Study of Villa Vannucchi in San Giorgio a Cremano (Italy). *Land*, *11*(9), 1568	p. 1	In this perspective, the circular city model represents a new way of organizing the city. As demonstrated by a variety of best practices, the entry points for triggering circular processes at the urban scale are various

(continued)

(continued)

No	References	Page	Definition
25	Coskun, A., Metta, J., Bakırlıoğlu, Y., Çay, D., & Bachus, K. (2022). Make it a circular city: Experiences and challenges from European cities striving for sustainability through promoting circular making. Resources, Conservation and Recycling, 185, 106495	p. 2	Hence, maintaining the community and sustaining the engagement of makers are important elements for realizing the idea of a circular city, a city that adopted CE to its core production and consumption systems
26	Tsui, T., Derumigny, A., Peck, D., Van Timmeren, A., & Wandl, A. (2022). Spatial clustering of waste reuse in a circular economy. Frontiers in Built Environment, 154	p. 2	The study of circular cities requires a spatial or geographical perspective. For cities, transitioning to a circular economy requires the introduction of circular (industrial) activities into the region, such as recycling, remanufacturing, storage, and (reverse) logistics; which are affected by spatial factors such as proximity to materials, clients, suppliers, and other companies
27	Horn, E., & Proksch, G. (2022). Symbiotic and regenerative sustainability frameworks: moving towards circular city implementation. Frontiers in Built Environment, 7, 178	p. 2	The circular city (CC) framework has become part of a larger transitional dialogue which envisions regenerative circularity and symbiotic resource flows across scales and contexts. Implementing circular principles in cities can involve actions such as scaling up integrated networks, retrofitting existing businesses, and creating new operational practices across scales. In doing so, challenges often exist, and there is a need for new tools, innovation, and approaches to future planning (Baganz et al., 2020). Achieving successful circular cities may be particularly contingent upon a successful transition toward reframing externalities to account for the true invaluable nature of ecosystems health and human wellbeing within economic and urban systems, something which has become all the more evident in light of the COVID-19 pandemic. Likewise, changing societal views and behaviors toward more sustainable practices, which are already in motion, are needed to strengthen a driving and supporting force for change

(continued)

(continued)

No	References	Page	Definition
28	Moraes, F. T. F., Gonçalves, A. T. T., Lima, J. P., & da Silva Lima, R. (2022). Transitioning towards a sustainable circular city: How to evaluate and improve urban solid waste management in Brazil. Waste Management & Research, 0734242X221142227	p. 1	A circular city (CC) is dedicated to closing loops and transforming linear processes into circular processes to eliminate waste and waste generation (Girard and Nocca, 2019). According to the Circular City Declaration (2020), a CC promotes business models and economic behaviours that decouple resources from economic activities, maintaining product value and utility and maintaining components, materials and nutrients for as long as possible. Consequently, cities can close material cycles and minimize harmful resource use and waste generation

References

1. Aksnes DW, Sivertsen G (2019) A criteria-based assessment of the coverage of scopus and web of science. J Data Inf Sci 4:1–21. https://doi.org/10.2478/jdis-2019-0001
2. Bîrgovan AL, Lakatos ES, Szilagyi A, Cioca LI, Pacurariu RL, Ciobanu G, Rada EC (2022) How should we measure? A review of circular cities indicators. Int J Environ Res Public Health 19(9):5177. https://doi.org/10.3390/ijerph19095177
3. Bocken NM, De Pauw I, Bakker C, Van Der Grinten B (2016) Product design and business model strategies for a circular economy. J Ind Prod Eng 33(5):308–320. https://doi.org/10.1080/21681015.2016.1172124
4. Bond, A. J., Dockerty, T., Lovett, A., Riche, A. B., Haughton, A. J., Bohan, D. A., Sage, R. B., Shield, I. F., Finch, J. W., Turner, M. M., & Karp, A. (2011). Learning how to deal with values, frames and governance in sustainability appraisal. *Regional Studies, 45*(8), 1157–1170. https://doi.org/10.1080/00343404.2010.485181
5. Campbell-Johnston K, ten Cate J, Elfering-Petrovic M, Gupta J (2019) City level circular transitions: Barriers and limits in Amsterdam, Utrecht and The Hague. J Clean Prod 235:1232–1239. https://doi.org/10.1016/j.jclepro.2019.06.106
6. Creswell, J. W. (2004). Educational research: Planning, conducting, and evaluating quantitative and qualitative research. Upper Saddle River, NJ: Pearson Education.
7. Egidi G, Salvati L, Vinci S (2020) The long way to tipperary: City size and worldwide urban population trends, 1950–2030. Sustain Cities Soc 60. https://doi.org/10.1016/j.scs.2020.102148
8. Ellen MacArthur Foundation. (2020). What is the circular economy? In Ellen MacArthur Foundation ANBI. (Accessed on Jan 20th 2023) https://www.ellenmacarthurfoundation.org/circular-economy/what-is-the-circular-economy
9. Gaustad G, Krystofik M, Bustamante M, Badami K (2018) Circular economy strategies for mitigating critical material supply issues. Resour Conserv Recycl 135:24–33. https://doi.org/10.1016/j.resconrec.2017.08.002
10. Geissdoerfer M, Savaget P, Bocken NM, Hultink EJ (2017) The Circular Economy–A new sustainability paradigm? J Clean Prod 143:757–768. https://doi.org/10.1016/j.jclepro.2016.12.048

11. Gomes GM, Moreira N, Ometto AR (2022) Role of consumer mindsets, behaviour, and influencing factors in circular consumption systems: A systematic review. Sustainable Production and Consumption. https://doi.org/10.1016/j.spc.2022.04.005
12. Gravagnuolo A, Angrisano M, Fusco Girard L (2019) Circular economy strategies in eight historic port cities: Criteria and indicators towards a circular city assessment framework. Sustainability 11(13):3512. https://doi.org/10.3390/su11133512
13. Hoornweg D, Bhada-Tata P, Kennedy C (2013) Environment: Waste production must peak this century. Nature 502:615–617. https://doi.org/10.1038/502615a
14. Kirchherr J, Reike D, Hekkert M (2017) Conceptualizing the circular economy: An analysis of 114 definitions. Resour Conserv Recycl 127:221–232. https://doi.org/10.1016/j.resconrec.2017.09.005
15. Konietzko J, Bocken N, Hultink EJ (2020) Circular ecosystem innovation: An initial set of principles. J Clean Prod 253:119942. https://doi.org/10.1016/j.jclepro.2019.119942
16. Korhonen J, Honkasalo A, Seppälä J (2018) Circular Economy: The Concept and its Limitations. Ecol Econ 143:37–46. https://doi.org/10.1016/j.ecolecon.2017.06.041
17. Korsunova A, Horn S, Vainio A (2021) Understanding circular economy in everyday life: Perceptions of young adults in the Finnish context. Sustainable Production and Consumption 26:759–769. https://doi.org/10.1016/j.spc.2020.12.038
18. Liang D, De Jong M, Schraven D, Wang L (2021) Mapping key features and dimensions of the inclusive city: a systematic bibliometric analysis and literature study. Int J Sust Dev World:1–20. https://doi.org/10.1080/13504509.2021.1911873
19. Liu Z, Schraven D, de Jong M, Hertogh M (2023) The societal strength of transition: a critical review of the circular economy through the lens of inclusion. Int J Sust Dev World 1–24. https://doi.org/10.1080/13504509.2023.2208547
20. Paiho, S., Mäki, E., Wessberg, N., Paavola, M., Tuominen, P., Antikainen, M., Heikkilä, J., Rozado, C. A., & Jung, N. (2020). Towards circular cities—Conceptualizing core aspects. Sustain Cities Soc 59:102143. https://doi.org/10.1016/j.scs.2020.102143
21. Paoli F, Pirlone F, Spadaro I (2022) Indicators for the Circular City: A Review and a Proposal. Sustainability 14(19):11848
22. Phulwani PR, Kumar D, Goyal P (2020) A Systematic Literature Review and Bibliometric Analysis of Recycling Behavior. J Glob Mark 33:354–376. https://doi.org/10.1080/08911762.2020.1765444
23. Potting, J., Hekkert, M. P., Worrell, E., & Hanemaaijer, A. (2017). Circular economy: measuring innovation in the product chain. *Planbureau voor de Leefomgeving*, (2544). (Accessed on May 20th 2023) https://www.pbl.nl/sites/default/files/downloads/pbl-2016-circular-economy-measuring-innovation-in-product-chains-2544.pdf
24. Rootes C (2009) Environmental movements, waste and waste infrastructure: an introduction. Environmental Politics 18(6):817–834. https://doi.org/10.1080/09644010903345587
25. Schraven D, Joss S, De Jong M (2021) Past, present, future: Engagement with sustainable urban development through 35 city labels in the scientific literature 1990–2019. J Clean Prod 292:125924. https://doi.org/10.1016/j.jclepro.2021.125924
26. Stahel WR, Clift R (2016) Stocks and flows in the performance economy. Taking Stock of Industrial Ecology 137–158. https://doi.org/10.1007/978-3-319-20571-7_7
27. Thomas DR (2006) A general inductive approach for analyzing qualitative evaluation data. Am J Eval 27(2):237–246. https://doi.org/10.1177/1098214005283748
28. Tsui, T., Derumigny, A., Peck, D., Van Timmeren, A., & Wandl, A. (2022). Spatial clustering of waste reuse in a circular economy. Frontiers in Built Environment, 154.
29. Vanhuyse F, Fejzić E, Ddiba D, Henrysson M (2021) The lack of social impact considerations in transitioning towards urban circular economies: a scoping review. Sustain Cities Soc 75:103394
30. Velenturf AP, Purnell P (2021) Principles for a sustainable circular economy. Sustainable Production and Consumption 27:1437–1457. https://doi.org/10.1016/j.spc.2021.02.018
31. Williams J (2019) Circular cities. Urban Studies 56(13):2746–2762. https://doi.org/10.1177/0042098018806133

32. Williams J (2021) Circular cities: what are the benefits of circular development? Sustainability 13(10):5725

Open Access This chapter is licensed under the terms of the Creative Commons Attribution 4.0 International License (http://creativecommons.org/licenses/by/4.0/), which permits use, sharing, adaptation, distribution and reproduction in any medium or format, as long as you give appropriate credit to the original author(s) and the source, provide a link to the Creative Commons license and indicate if changes were made.

The images or other third party material in this chapter are included in the chapter's Creative Commons license, unless indicated otherwise in a credit line to the material. If material is not included in the chapter's Creative Commons license and your intended use is not permitted by statutory regulation or exceeds the permitted use, you will need to obtain permission directly from the copyright holder.

Exploring the Inclusive City: Definitions and Dimensions

Danni Liang, Martin de Jong, and Daan Schraven

Abstract The political and public interest in issues of inclusion and inclusiveness has grown steadily in recent years. Keeping different segments of society together in the aftermath of a neo-liberal era where much of the social tissue underlying market operations has been eaten up by the prevalence of those same market values is a key concern to many public and private actors. The popularity of the label 'inclusive city' can also be observed in its increased use among municipal governments worldwide for city branding purposes and its surge in the academic literature. Its relevance notwithstanding, the meaning of the term 'inclusive' is not always clearly defined and often multi-dimensional. In this chapter, a state-of-the art overview will be offered of what is currently known about this city label in the academic literature and look both at journal articles and books in the timeframe 2000–2022. Key finding in this study, which builds on and further develops earlier work is that based on both bibliometric research of academic articles and a systematic review of books, book chapters and grey literature, we find six different dimensions of inclusion (spatial, social, environmental, economic, political and cultural) with their own connotations and associations. Taking this variety into account is essential to a more sophisticated understanding of what developing an inclusive city entails and what variations and variety of developmental paths exist.

D. Liang (✉)
School of Law, Wenzhou University, Wenzhou, China
e-mail: liangdn@wzu.edu.cn

M. de Jong
Rotterdam School of Management and Erasmus School of Law, Erasmus University Rotterdam, Rotterdam, The Netherlands

Institute for Global Public Policy, Fudan University, Shanghai, China

M. de Jong
e-mail: w.m.jong@law.eur.nl

D. Schraven
Faculty of Architecture in the Built Environment, TU Delft, Delft, The Netherlands
e-mail: D.F.J.Schraven@tudelft.nl

M. de Jong
Smart City Institute, HEC-Liege, University of Liege, Liege, Belgium

Keywords Inclusive city · Inclusion · Bibliometric study · Qualitative literature review · Dimensions of the inclusive city

1 Introduction

Municipalities increasingly brand themselves as 'inclusive', sometimes as just that and sometimes in combination with other attractive labels such as 'inclusive smart' or 'sustainable and inclusive'. This same trend can be observed in bibliometric studies that show a surge in its use in recent years, although as such the 'smart city' and the 'sustainable city' remain the most popular categories by far [56].

Growing attention paid to 'inclusion' and 'inclusiveness' is not unique to the fields of urban studies and environmental policy; it has become of key importance in a great variety of fields and can be seen as the almost logical result of decades in which market domination, shareholder value, economic growth and allocative efficiency prevailed over government intervention, balancing stakeholder interests, quality of life more broadly defined and distributional effects. Maintaining in place what are often called 'neo-liberal policies' as the dominant political and administrative ideology in most countries worldwide has gone at the expense of the solidity and solidarity of social tissue. The consequences of these policies have become apparent in many cities, especially in Anglo-Saxon countries where their influence was at its strongest, but also elsewhere. Public infrastructures suffer from lack of proper upkeep, housing prices have gone through the roof leading to the dramatic appearance of homelessness, cleanliness of public space leaves to be desired and segregation has led to growing interethnic tensions. More generally, the gap between haves and have-nots has grown quite significantly and substantial portions of the underprivileged struggle to find decently paid employment or otherwise have difficulties to get by in times of high inflation. Exclusion has thus become an undeniable phenomenon, but that does not yet answer the question which categories or types of people are excluded, and from which benefits, facilities or privileges exactly they are excluded.

It is the aim of this chapter to address that last topic by examining through both a quantitative and a qualitative literature review what leading authors in the field have written on it and how their findings can be systematized and classified. Some authors have made dedicated efforts to establish conceptual frameworks of inclusive urban development by means of quantitative approaches and mathematical models; others have placed particular emphasis on different areas within the broader field of sustainable urban development, such as inclusive economic growth, spatial accessibility, cultural diversity and social cohesion; yet another group has focused on policy outcomes and policy impacts from the perspective of public policy and policy analysis. In spite of the insights offered by abovementioned contributions, knowledge of how the inclusive city can be defined, what various dimensions it has and how it can be realized have not been systematically examined.

Knowledge on the concept of the inclusive city is fragmented, which may well lead to high but unfounded expectations or ill-guided policy actions. That knowledge

Exploring the Inclusive City: Definitions and Dimensions

gap can be seen as the starting point for this contribution. One can only develop an inclusive city if one knows what it is, what aspects there are to it and through what policy actions these can be synthesized or traded off against each other. Below the main definitions of the concept 'inclusive city' will be mapped, the various conceptual dimensions explored and the interrelationships between relevant keywords related to the concept analysed.

In Sect. 2, we will briefly outline our methodological approach. In the third section of this chapter, the bibliometric method is used to tease out the various dimensions of the inclusive city and identify the interrelationships that exist between different keywords related to it. A qualitative analysis is then used in Sect. 4 to review the concept of the inclusive city and explore relevant definitions discerned in it. Finally, concluding Sect. 5 takes stock of the findings in the previous sections and maps them and synthesises them into a graphical display with six different dimensions of the inclusive city and their respective policy-relevant connotations.

2 Methodological Approach

2.1 Research Design

This survey into the concept 'inclusive city' leans strongly on a previous study with a specific research framework presented earlier in [40]. However, it provides an update and upgrade of it, not only because its bibliometric study and qualitative literature survey include the years 2021 and 2022, but in addition to that, recent shifts in attention to aspects of the inclusive city are also discussed at length. The bibliometric part of the analysis evolved in three steps:

- High-frequency keywords in the field of inclusive cities were counted and analysed so that key research contents and topics of this field were obtained.
- A co-occurrence analysis was conducted to reveal the interrelationships between high-frequency keywords within the research domain of inclusive cities.
- A cluster analysis was performed using SPSS to identify clusters in the field and to further deepen the conceptual underpinnings of the concept 'inclusive city'.

Following that, the bibliometric analysis was complemented with a qualitative review of academic books, policy reports, lecture notes and grey literature, to further deconstruct the concept and gain additional theoretical knowledge of it (Fig. 1).

2.2 Data Collection

The research design started with data collection for the bibliometric analysis. In our search strategy, the focus was on the inclusive city concept as a vehicle of sustainable urban development. As a novel concept, 'inclusive city' could be recognized in the title, abstract and as an author keyword as part of the academic literature. In order to systematically capture the relevant research on inclusive cities, we collected bibliometric data on articles using the following search query:

> TITLE-ABS (*"inclusive city"* OR *"inclusive cities"*) OR AUTHKEY (*"inclusive city"* OR *"inclusive cities"*) AND DOCTYPE (ar OR re) AND PUBYEAR < 2023 AND PUBYEAR > 1999 AND LANGUAGE (English)

A few implicit decisions were made with regards to this query:

First, Scopus was used to compile the library of academic articles referring to inclusive city as a concept. For one thing, Scopus indexes a larger number of journals than Web of Science, and includes more international and open access journals [4]. Also, Scopus fits the aim of this study because of its comprehensiveness in covering a wide range of journals, thereby ring-fencing a multitude of possible dimensions that the inclusive city could target.

Second, the analysis was centred on academic journal articles and reviews in the English language within the timeframe 2000–2022. Academic papers and reviews offer a stable, verified and accessible account of the academic literature, which helps to initially profile and subsequently review the different angles of attention to the inclusive city through key words. The longitudinal scale of the data sample allowed for exploration of the knowledge of the inclusive city concept. Title, abstract and author keywords are three bibliometric locations that convey the essence of a published study and therefore largely reflect the position of the inclusive city within relevant research fields. Based on the search strategy and the above criteria, 184 publications from Scopus database were finally retrieved for subsequent bibliometric analysis.

The next step aimed to uncover the definitions and conceptual meaning of the inclusive city concept: a qualitative survey independent from the bibliometric analysis. Books, book chapters and valuable reports provide more clarity on conceptual underpinnings of the inclusive city, which are different from the documents used in the bibliometric analysis. We collected books and book chapters in the English language in the Scopus database from 2000 to 2022 with the following search query:

> TITLE-ABS ("inclusive city" OR "inclusive cities") OR AUTHKEY ("inclusive city" OR "inclusive cities") AND DOCTYPE (bk OR ch) AND PUBYEAR < 2023 AND LANGUAGE (English)

Books and book chapters were selected for in-depth review, if these were cited at least more than once. In this way, the input of these sources could at least be assumed to have some academic resonance. Relevant books on the topic 'inclusive city' through Amazon Books were also collected as a data source of this study.

Additionally, we collected a selection of valuable reports from the official websites of leading international organizations and institutions (i.e., United Nations, the World Bank, OECD and Asian Development Bank) for qualitative analysis. A total of 20 books, 11 book chapters and 11 reports were selected for further inspection (see Table 1). When listing the sources in Table 1, we followed the order of their year of publication first, and then the order of the document types.

Table 1 Overview of selected books, book chapters and reports for qualitative review

Author	Year	Title	Type
D. Westendorff	2004	From Unsustainable to Inclusive Cities	Book
P. Herrle, U. Walther	2005	Socially Inclusive Cities: Emerging Concepts and Practice	Book
A. Laquian, L. Hanley	2007	The Inclusive City: Infrastructure and Public Services for the Urban Poor in Asia	Book
F. Steinberg, M. Lindfield	2011	Inclusive Cities	Book
C. Whitzman, C. Legacy, C. Andrew et al	2013	Building Inclusive Cities: Women's Safety and the Right to the City	Book
R. Hambleton	2014	Leading the Inclusive City: Place-based Innovation for a Bounded Planet	Book
J. Gupta, K. Pfeffer, H. Verrest et al	2015	Geographies of Urban Governance: Advanced Theories, Methods and Practices	Book
N. Espino	2015	Building the Inclusive City: Theory and Practice for Confronting Urban Segregation	Book
S. Venkateswar, S. Bandyopadhyay	2016	Globalisation and the Challenges of Development in Contemporary India (Dynamics of Asian Development)	Book
D. Zuberi, A. Taylor	2017	(Re)Generating Inclusive Cities: Poverty and Planning in Urban North America	Book
S. Attia, Z. Shafik, A. Ibrahim	2018	New Cities and Community Extensions in Egypt and the Middle East: Visions and Challenges	Book
J. Salahub, M. Gottsbacher, J. de Boer	2018	Social Theories of Urban Violence in the Global South: Towards Safe and Inclusive Cities	Book
N. Pokhrel	2019	Transforming Kolkata: A Partnership for a More Sustainable, Inclusive, and Resilient City	Book
V. Bharne, S. Khandekar	2019	Affordable Housing: Inclusive Cities	Book
J. Salahub, M. Gottsbacher, J. De Boer et al	2019	Reducing Urban Violence in the Global South: Towards Safe and Inclusive Cities	Book
D. Kundu, R. Sietchiping, M. Kinyanjui	2020	Developing National Urban Policies: Ways Forward to Green and Smart Cities	Book
V. Pineda	2020	Building the Inclusive City: Governance, Access, and the Urban Transformation of Dubai	Book
B. Dahiya, A. Das	2020	New Urban Agenda in Asia–Pacific: Governance for Sustainable and Inclusive Cities	Book
A. Anttiroiko, M. De Jong	2020	The Inclusive City: The Theory and Practice of Creating Urban Prosperity for all	Book

(continued)

Table 1 (continued)

Author	Year	Title	Type
T. P. Uteng, H. R. Christensen, L. Levin	2020	Gendering Smart Mobilities	Book
K. Viswanath	2013	Gender Inclusive Cities Programme: Implementing Change for Women's Safety	Book chapter
C. Andrew, C. Legacy	2013	The Role of Partnerships in Creating Inclusive Cities	Book chapter
A. Schippers, L. Van Heumen	2014	The Inclusive City through the Lens of Quality of Life	Book chapter
A. Schippers, L. Van Heumen	2014	The inclusive city through the lens of quality of life	Book chapter
N. Sridharan	2015	Can Smart City Be an Inclusive City? - Spatial Targeting (ST) and Spatial Data Infrastructure (SDI)	Book chapter
V. Walters	2016	Urban Neoliberalism and the Right to Water and Sanitation for Bangalore's Poor	Book chapter
A. Morgan	2019	"Dad, Do Not Cry": Imagination and creativity on their own terms in inclusive cities and communities	Book chapter
V. R. Sharma, Chandrakanta	2019	Perspective on Resilient Cities: Introduction and Overview	Book chapter
D. Dahiya, A. Das	2020	New Urban Agenda in Asia–Pacific: Governance for Sustainable and Inclusive Cities	Book chapter
A. A. Popoola, N. V. Blamah, C. Mosima et al	2021	The Language of Struggle and Radical Activism as an Inclusive City Tool Among the Neglected Urban Poor of South Africa	Book chapter
R. Sultana, A. Asad	2021	Evaluation of Urbanites' Perception About Livable City Using Analytic Hierarchy Process (AHP): A Case Study of Dhaka City	Book chapter
United Nations Centre for Human Settlements	2001	The State of the World's Cities	Report
Asian Development Bank	2010	Access to Justice for the Urban Poor: Toward Inclusive Cities	Report
World Bank	2015	Inclusive Economic Growth in America's Cities: What's the Playbook and the Score?	Report
United Nations General Assembly	2015	Transforming our world: the 2030 Agenda for Sustainable Development	Report
World Bank	2015	World-Inclusive Cities Approach Paper	Report
UN-Habitat III	2015	Habitat-III-Issue-Paper-1_Inclusive-Cities	Report
OECD	2016	Making Cities Work for All: Data and Actions for Inclusive Growth	Report
UN-Habitat III	2017	The New urban agenda	Report
Asian Development Bank	2017	Enabling Inclusive Cities: Tool Kit for Inclusive Urban Development	Report

(continued)

Table 1 (continued)

Author	Year	Title	Type
United Nations	2020	The policy Guidelines for Inclusive Sustainable Development Goals	Report
Asian Development Bank	2022	Inclusive Cities-Urban Area Guidelines	Report

2.3 Methods

In order to explore the knowledge distribution structure in the inclusive city research domain and deepen our comprehension of the concept, we statistically analysed and summarized the number of high-frequency keywords, the frequency as well as the betweenness of each keyword in an article or a review. The results of the high-frequency analysis convey information about the variety in focus and the state of the inclusive city research field, where high frequency of occurrence and high betweenness of keywords can both indicate a high importance of the keywords. More specifically, the frequency of keywords is positively correlated with their research popularity. The betweenness of a keyword represents the strength of its connection to other keywords, meaning that higher betweenness implies more connections to keywords. In other words, the keywords act as a hub in the research field and play a "bridging" role in the development of research topics. Before the calculation, irrelevant and meaningless keywords were removed to make the results of the analysis more accurate and rigorous; and some keywords with similar academic meanings and relatively low frequency of occurrence (no more than 3 occurrences) were combined and renamed to avoid unexpected omissions in the summary of high-frequency keywords and potential misunderstanding. Table 2 shows the result of that procedure in which 25 renamed keywords were obtained.

Following the above *keyword co-occurrence analysis*, the frequency of two keywords simultaneously appearing in the same article was counted, thus revealing the correlation strength of different keywords in an article. The more frequently two keywords simultaneously appeared in the same document, the more explicitly the connection between the two keywords has been made. The size of nodes in the co-occurrence network is determined by the occurrences of keywords. A larger node reflects a higher correlation with the research topic: the inclusive city. In addition, the line between two nodes is called a link, indicating the strength of the co-occurrence relationships between different keywords. We used the visualisation software of VOSviewer to construct a co-occurrence network of keywords in order to present and reveal the interrelationships between different keywords [23].

During the next step, the analysis focused on exploring different clusters composed of closely linked high-frequency keywords by means of a *cluster analysis*, identifying the conceptual structure of the inclusive city concept. Cluster analysis is a way of grouping cases of data based on the similarity of responses to several variables, and its principle is to give the keywords within the same category as high a homogeneity as

Table 2 Complete list of merged and renamed keywords

Renamed keywords	Included original keywords	Reasons for combination
Accessibility	Accessibility; accessibility strategies in 2030;	Similar academic meaning
Environment	Environment; environments; built environment; environment equity; living environment; walking environment	
Governance	Urban governance; inclusive governance; land governance; governance; multi-level governance;	
Inclusion	Inclusion; inclusiveness; disability inclusion; social inclusion;	
Housing	Affordable housing; housing; social housing; housing policies; public housing; housing cooperatives	
Public space	Public space; public spaces; urban open space; open space;	
Rights	Rights; human rights; right to the city; empowerment; language rights; right to housing;	
Sustainability	Sustainability; sustainable; social sustainability;	
SDGs	sdg; sdgs; sdgs 11 & 10.2;	Differences in singular and plural forms, but similarities in academic meaning
Urbanization	Urbanization; urbanisation;	
Inclusive city	Inclusive city; inclusive cities;	
Smart city	Smart city; smart cities;	
Neighbourhood	Neighbourhood; neighbourhoods;	
Planning	Planning; participatory-collaborative planning; modern planning; urban planning; city planning; planning interventions; planning and design; green-oriented urban planning; land use planning; citizen-centric urban planning; smart urban planning; technology-aided urban planning	Similar academic meaning and relatively low frequency of occurrences (no more than 3 occurrences)
Community	Community art; community capital; community organizations; community savings; community-based organizations; gated communities;	
Migration	Migratory phenomena; migration; international migration;	
Citizenship	Citizenship; citizens' perception; citizen–state relations;	

(continued)

Table 2 (continued)

Renamed keywords	Included original keywords	Reasons for combination
Participation	Participation; participatory process;	
Transport	Green transportation; transport inequality; transport policy; sustainable transport; commuting burden; tramway; traffic congestion; railway; congestion tax;	
Economic regeneration	Economic regeneration; informal economy; employment; local economy	
Infrastructure	Urban green infrastructure; urban infrastructure; inclusive infrastructure; living infrastructure; urban green infrastructure; soft infrastructure; smart infrastructure; infrastructure of mobility;	
Segregation	Urban segregation; socio-spatial segregation; segregation;	
Finance	Innovative finance; alternative finance; local finance; municipal finance; housing finance;	
Engagement	Social engagement; community engagement; engagement channels;	
Mobility	Mobility; mobility of care; mobility planning; independent mobility; children's independent mobility; urban mobility	

possible, while the heterogeneity between categories is as high as possible. Therefore, the aim of using cluster analysis in this study was to identify groups of keywords, that are connected based on a stronger association with each other than to other keywords from other clusters. In this article, hierarchical clustering was utilized because it can help uncover various aspects of the inclusive city concept when it is not clear how many clusters are distinguished, and it is an aid in exploring the hierarchical relationship between clusters. Ward's method was used to join cases into clusters such that the variance within a cluster was minimized. A more detailed explanation of this application can be found in in [40].

3 Quantitative Findings

3.1 Publication Activities in the Inclusive City Literature

Analysing numerical changes in the literature on inclusive cities through the years helps to portray overall trends in the field as well as conceptual hot spots within the domain. To assess the basic publishing activities in the inclusive city literature and their evolution over time, the annual number of publications in our dataset in Scopus from 2000 to 2022 was first counted (see Fig. 2). We can see from Fig. 2 that the annual numbers of publications in this research field generally show a trend of slow fluctuation followed by rapid growth over time. The number of publications has been steadily increasing, which reflects widespread interest among scholars in studies related to the inclusive city. Specifically, although the quantity of annual publications was overall relatively small, it increased slightly between 2000 and 2016 (reaching an incidental small peak in 2010) and then saw a steep incline from 2016 on. In terms of the volume and incremental growth rates, the rise in the period 2010–2017 and even more so in 2018–2022 turns out to several times the previous increase. It is obvious that research on the inclusive city has received growing attention in recent years and as a research topic it has acquired strong momentum.

Figure 3 clearly shows the distribution across different disciplines relevant to research on the inclusive city from 2000 to 2022. With the emergence of new research findings in recent years, scholars from different disciplines have begun to introduce

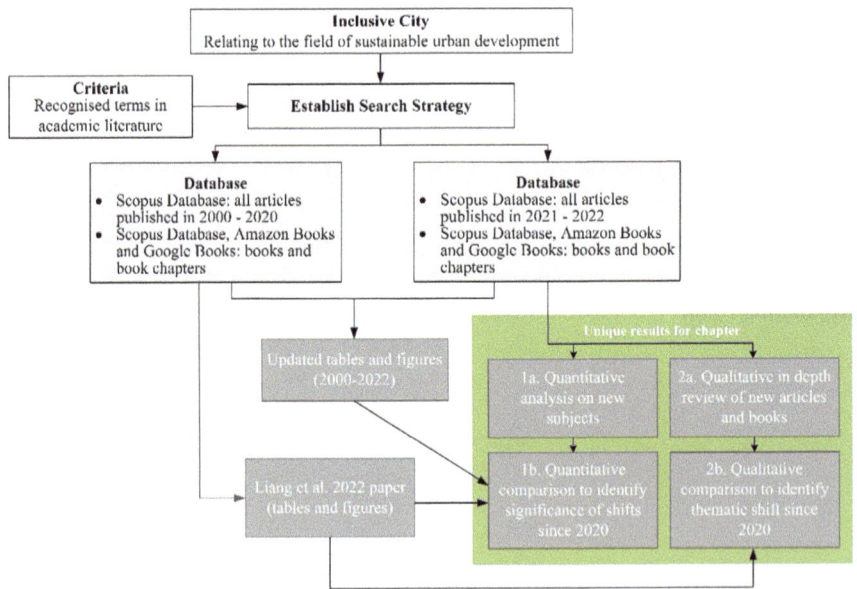

Fig. 1 Research design

Exploring the Inclusive City: Definitions and Dimensions

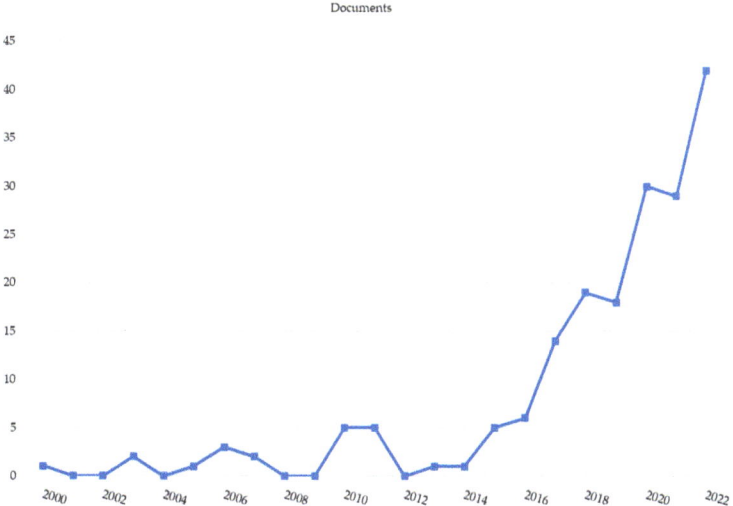

Fig. 2 Total number of publications about inclusive cities research (2000–2022)

multiple perspectives of the inclusive city. Clearly, the largest number of publications on inclusive cities appeared in the category social sciences, with 83.70%, followed by environmental science, with 34.78%, which is general agreement with [40]. The difference is that the number of publications on inclusive cities in the discipline of engineering has now reached the third highest level. It is noteworthy that the sum of the publications in each discipline is greater than 100%, indicating that inclusive cities are a multidisciplinary and cross-cutting research field.

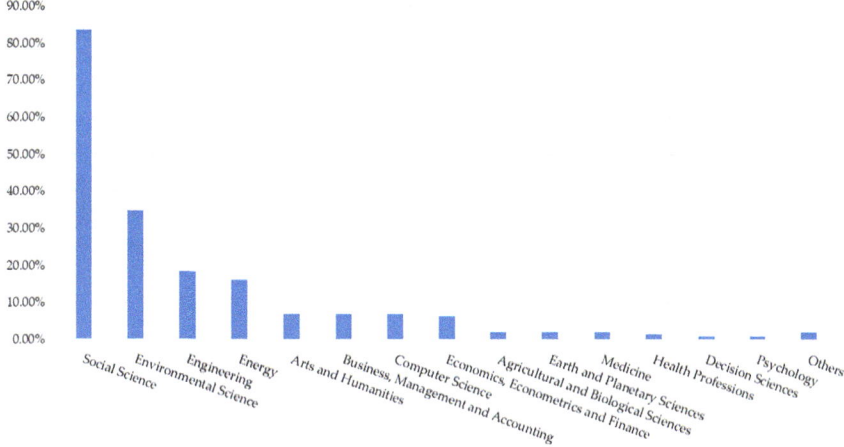

Fig. 3 Disciplines and their publication numbers regarding the inclusive city (2000–2022)

Table 3 High-frequency keywords related to the inclusive city

Keyword	Frequency	Betweenness	Keyword	Frequency	Betweenness
Inclusive city	45	0.20	Disability	7	0.00
Planning	22	0.04	Infrastructure	7	0.00
Inclusion	20	0.06	South Africa	6	0.01
Governance	16	0.11	Environment	6	0.03
Smart city	15	0.04	Segregation	6	0.00
Public space	13	0.03	Mobility	6	0.01
Sustainability	13	0.04	Community	6	0.01
Accessibility	12	0.01	Innovation	5	0.01
Housing	11	0.04	Resilience	5	0.01
Migration	11	0.01	Finance	5	0.01
Informal	11	0.04	Economic regeneration	5	0.00
Participation	11	0.03	Urbanization	5	0.00
Rights	10	0.02	Gender	5	0.00
Transport	10	0.00	Urban regeneration	5	0.00
Land use	8	0.01	engagement	5	0.01
SDGs	7	0.01	India	4	0.00

The reason why we chose high-frequency keywords in the following steps is that they represent a high concentration and the core content in the literature and offer an indication of the direction in which this emerging research domain is moving. The keywords in Table 3 represent the topics that received the most attention in articles about the inclusive city. Among them, keywords such as planning, inclusion, governance, smart city, public space, sustainability and accessibility, show higher frequencies of occurrence and stronger levels of co-occurrence and thus occupy the more central positions in the network of relations. In contrast to [40], some new high-frequency keywords have emerged, such as disability, resilience, urbanization, gender and urban regeneration, indicating that the above topics have gradually become new research hotspots in recent years and providing fresh additional perspectives to make sense of the inclusive city concept. Overall, these keywords reflect the meaning of inclusive cities as well as core issues and insights in the research domain. However, a richer picture can be obtained if their underlying relationships are further explored with co-occurrence analysis.

3.2 Co-occurrence of High-Frequency Keywords

In order to examine the relationships between these high-frequency keywords related to inclusive cities, we established a co-occurrence network displayed in Fig. 4. We can see from it that "inclusive city" itself had the largest node, followed by

Exploring the Inclusive City: Definitions and Dimensions

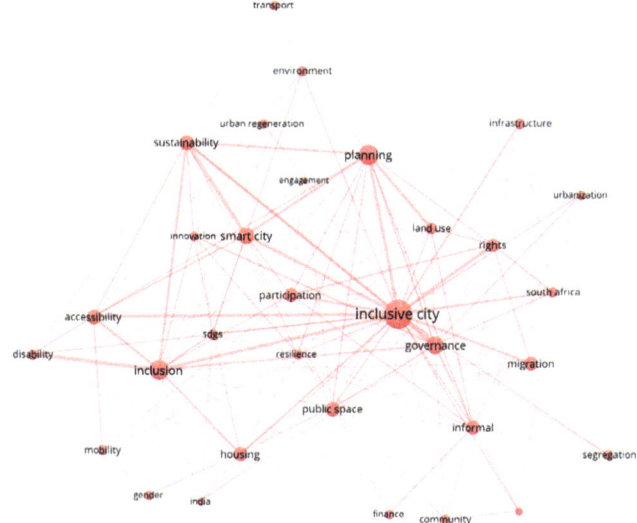

Fig. 4 Co-occurrence network of high-frequency keywords

"planning", "governance", "inclusion", "sustainability", "smart city", "accessibility","public space", "rights" and "migration". It should be noted that "inclusive city" is the central node in the network and that it has close connections with "governance", "sustainability", "planning" and "public space". In addition, "sustainability", "planning" and "governance" are seen as research focuses that play a vital role in the inclusive cities research domain. On the whole, the high-frequency keywords in the field are strongly linked and correlated, indicating that the research hotspots have grown increasingly focused and that the research field has been strengthened in scope, relevance and depth.

Closely related high-frequency keywords can be associated by cluster analysis so as to form various classes and show the structure of relevant topics in the research field. As output of the cluster analysis, we developed a tree diagram demonstrating the structure and relationships between different keywords (see Fig. 5). Each case began as a cluster and subsequently the two most similar cases were found (e.g., urbanization and land use) by looking at the square Euclidean distances between pairs of cases. The next case merged was the one with the highest similarity to urbanization or land use, and so on (e.g., migration, South Africa, rights). Finally, all high-frequency keywords could be divided into seven topics at the threshold of 10. However, the clustering results obtained in this study differ from those in [40]. Table 4 shows a further comparison of the clustering results between the two studies.

Our 1st cluster has been labelled "Migration and rights" as it included the following keywords: "land use, South Africa, inclusive city, migration, rights, urbanization", reflecting the core content of both the clusters of "Space and rights" and "Sustainable migration" obtained in [40]. The 2nd cluster was named "Segregation and economic regeneration" as it contained the following keywords: "community,

Fig. 5 The tree diagram of cluster analysis

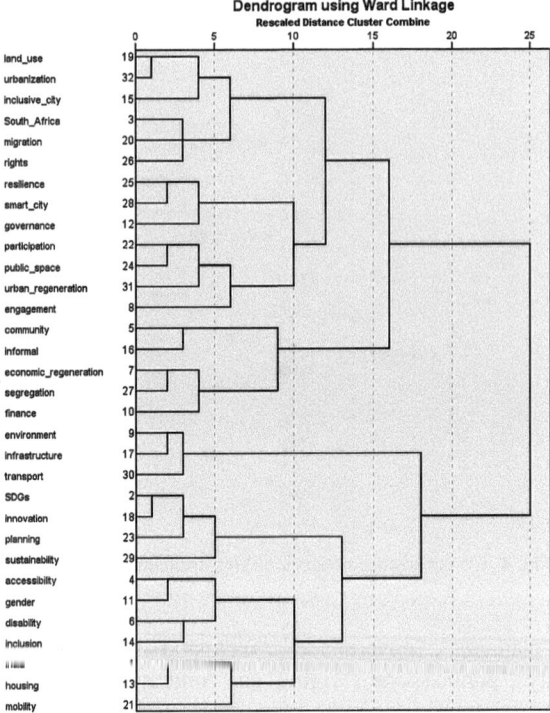

finance, informal, economic regeneration, segregation", indicating the intension of the two clusters named "Community and finance" and "Segregation and economic regeneration" obtained in [40]. The 3rd cluster was entitled "Smart governance and urban resilience" since it contained the following keywords: "governance", "engagement", "public space", "smart city", "resilience", "participation" and "urban regeneration". Our 4th cluster can be summarized as "Infrastructure and environment" which is consistent with the clustering results of [40], it included as keywords: "transport", "environment", "infrastructure". The 5th cluster received the denomination "Sustainable innovation and planning" as it included four keywords: "SDGs", "innovation", "planning", "sustainability". The 6th cluster was named "Accessibility and inclusion" as it included the following keywords: "accessibility", "disability" and "gender inclusion". Finally, the last cluster became known as "Mobility and housing" as it contained the three keywords: "India", "housing" and "mobility".

Following the size of the nodes, the strength of the correlation shown in Fig. 4 and the comparison between the clustering results of the two studies shown in Table 4, the main content of each of seven clusters can be outlined. The clusters can be described as follows:

Cluster 1 (Migration and rights): Rights, migration and land use in the process of urbanization are all closely related to the inclusive city. In recent years, migration has become a global phenomenon [60]. Particularly in South Africa, factors such as

Table 4 Comparison of the clustering results of the two studies

Topics (2000–2022)	Keywords	Topics (2000–2020)	Keywords
Migration and rights	Land use, South Africa, inclusive city, migration, rights, urbanization [**urbanization is an addition to** [40]]	• Space and rights	Public space, housing, rights, land use
		• Sustainable migration	Migration, sustainable development
Segregation and economic regeneration	Community, finance, informal, economic regeneration, segregation	• Community and finance	Community, finance, informal
		• Segregation and economic regeneration	segregation, South Africa, inclusive city, economic regeneration
Smart governance and urban resilience	Governance, engagement, resilience, public space, participation, smart city, urban regeneration [**resilience, urban regeneration are additions to** [40]]	Smart participation and citizenship	Smart city, accessibility, participation, mobility, engagement, citizenship
Infrastructure and environment	Transport, infrastructure, environment	Infrastructure and environment	Planning, infrastructure, transport, environment
Sustainable innovation and planning	SDGs, innovation, planning, sustainability	Sustainable innovation and governance	Governance, innovation, sustainability, inclusion, SDGs
Accessibility of women and the disabled	Accessibility, disability, gender, inclusion [**all are new additions to** [40]]		
Space and mobility	India, housing, mobility [**all are new additions to** [40]]		

nationality, race, gender and language frequently act as barriers that prevent immigrants from accessing a variety of public services and resources, which is closely linked to its history of immigration and its unprecedented urbanization [25]. In addition, the land use planning environment in South Africa is plagued by complex challenges, mainly due to weak enforcement provisions, low levels of participation as well as complexities regarding entitlements and claims on land [50]. Considering the social complexities involved, these processes are now mainly driven by experts. Instead, ensuring that the diverse rights, interests, and claims of broader stakeholder groups are adequately addressed should be prioritized. In addition, judicial vindication of the right to use land can provide impoverished and vulnerable populations to gain a foothold in urban areas, which is essential for South Africa's sustainable development.

Cluster 2 (Segregation and economic regeneration): Different from [40], this cluster includes a greater variety of aspects. In fact, segregation is considered a major obstacle to the creation of inclusive cities, while exclusion at the economic level is a serious threat to growth in urban prosperity, citizen well-being and social stability. In the post-pandemic era, global cities face specific problems in need of resolution, such as shortage of food production, low levels of local economic development and job creation for marginalized populations and high vacancy rates among commercial buildings [43]. Many community-based financing initiatives are good examples. In practice, community finance has provided support to enable individuals and organizations in creating wealth among disadvantaged communities and support them in acquiring better access to credit facilities, which indirectly affects their socioeconomic and political status [2, 35, 36].

Cluster 3 (Smart governance and urban resilience): Urban regeneration and resilience have become rising topics of interest in research on inclusive cities in the last two years [3, 41, 48]. The damage caused by COVID-19 to the natural and human environments as well as to social and economic development in cities worldwide has forced all to consider how cities can recover, thereby achieving more liveable and inclusive cities. This will largely depend on existing policies, planning, finance, digital infrastructure and governance systems [70]. Smartness in governance is actually reflected in the use of technology and data in ways that promote public participation, urban services and urban design, resulting in more efficient and effective decision-making. This includes ensuring the participation and collaboration of different stakeholders in the design of urban space [63]. This makes it easier for citizens to communicate with local governments through continuous upgrades in data systems and platforms and provides early prediction and warnings of social risks by collecting monitoring data from various sectors. In this regard, smart and digital solutions reflect the need for governance.

Cluster 4 (Infrastructure and environment): The role of the physical environment and infrastructure in promoting intergroup social inclusion has drawn a lot of attention. Well-designed urban neighbourhoods, when equipped with streets, squares, parks, markets, public transport and other infrastructures, provide abundant opportunities for public encounters among people from different backgrounds [27, 33, 71]. In the past two years, climate change and rapid urbanization have exacerbated global warming, leading to more extreme weather events. Major changes in the urban environment have increased the thermal vulnerability of urban residents, and different urban infrastructures enable different types of response to it. More proactive interventions have been proposed, such as early warning systems for the elderly and community-specific green infrastructure programmes [32].

Cluster 5 (Sustainable innovation and planning): The link between sustainability and innovation is often two-way: innovation can be considered a means to realize sustainability, just as sustainability can be described as a main purpose for innovation in urban areas. The primary aim of sustainable innovation is to significantly reduce negative environmental impact and promote sustainability. As an effective green governance model, it is applied to explore the possibility of strengthening environmental inclusion through innovative models, methods and technologies to ease the tension between expansive human aspirations and the deteriorating natural environment [39].

Cluster 6 (Accessibility of women and the disabled): There has been a significant increase in attention to the inclusion of disability and gender in the past two years. Despite a number of initiatives implemented by relevant institutions to support the disabled in many countries, this segment of the population still faces persistent barriers across all areas of urban life [30]. This is largely due to a lack of in-depth dialogue among policy makers and the community of the disabled or a lack of understanding regarding the interaction between disabled people and the built environment [64]. In the context of rising numbers of people with disabilities worldwide, the shift from traditional decision-making by major institutions to participatory consultation within communities has also rapidly promoted their inclusion. In addition, a gendered perspective is needed for inclusive and smart cities [18]. The importance of listening to the voices of those concerned and understanding what makes for a livable city to them is widely acknowledged in academic studies [58]. Greater understanding of how women and people with disabilities gain access (or not) to public space, infrastructures, and services provided in/by them is critical to boosting equality for them.

Cluster 7 (Space and mobility): Exclusion from housing and motorized mobility were particular acute during the COVID-19 pandemic. In many developing countries, challenges to access poor women have to housing not only remain, but they have in fact been exacerbated [1]. In other words, those gender-blind interventions have failed to take cognizance of the gendered impact of the pandemic on their housing experience (as was observed in South African cities, where women's exclusion from affordable housing during after COVID-19 became particularly pronounced [19]. Likewise, many studies emphasize the comprehension and configuration of mobility in relation to inclusion. Viewed from this perspective, promoting transit-oriented development, making investments in publicly funded transportation, and installing secure non-motorized infrastructure facilities are essential steps toward advancing mobility justice and building an inclusive city for all [45].

4 Qualitative Findings

Bibliometric findings provide an excellent opportunity to create a multidimensional depiction of the inclusive city, but certain questions, such as whether different perspectives lead to different understandings of the inclusive city are best answered through a qualitative review of the literature.

The term inclusive city was first promoted by the United Nations in 2001 and described as a place where everyone, regardless of their economic status, gender, race, ethnicity or religion, is enabled and empowered to fully participate in the social, economic and political opportunities that are on offer [66]. Subsequently, some key drivers of inclusive urban development aimed at putting people and their immediate needs at the forefront were proposed by UN-Habitat and the World Bank [65, 74]. This includes aspects such as political commitment, participation and social innovation, high-quality basic services, inclusive spatial planning, accountability and governance, financial and technical assistance and building partnerships. The OECD further provided a multidimensional framework for inclusion with two indicator domains: human and social capital (income, jobs and education) and urban environment (housing, transport, environment, safety, social support and subjective well-being) [46]. Notably, global commitment to sustainable urban development was reaffirmed through the adoption of the New Urban Agenda in 2016. Moreover, the Asian Development Bank developed an integrated framework with four critical aspects, accessibility, affordability, resilience and sustainability to describe specific characteristics of the inclusive city [59]. Based on the approved standard and international practices, the guidelines the ADB provided in 2022 focused on the importance of engaging in inclusive design solutions for people with disabilities, the elderly and children. These documents may not have been the most sophisticated analytical depictions of the inclusive city as a concept, but they were extremely instrumental in coining it as a crucial one for future policy initiatives.

The theory and practice of inclusive cities have also received growing attention in academic work. Discussions of exclusion and inclusion have largely focused on how social status and power are distributed unequally nowadays. As noted by [24], segregation and exclusion go hand in hand and create the setting in which inequality is reflected in urban space [24]. He expresses the opinion that "social disparities require physical segregation" which is at the heart of much of contemporary urban development. In other words, one's position in social space is believed to reflect one's level of wealth and status [24]. However, the interaction between people with different levels of wealth, power and social status as well as their mutual understanding have been reduced as a result of this spatial separation. Therefore, Espino makes it clear that differences in status and social class translate into a form of exclusion and that combating social and spatial segregation is key to creating a more equitable urban society.

Urban violence resulting from a lack of social inclusion can be seen as the other form of exclusion. More precisely, it is caused by political and social confrontations between different population groups and the destruction of ancient social norms that used to rule urban life, without these being substituted by new rules [17]. Curbing urban violence cannot depend on the wealth or poverty of the city but should be the result of the validity of the social pact, public policies, and societal norms. Spatial equality and social inclusion will only emerge from new behavioral norms and changes in institutional rules of the game for urban governance, preferably without resort to any forms of urban violence (i.e. gendered violence, state violence and interpersonal violence) [52, 53].

In addition to the two aspects mentioned above, exclusion is often frequently discussed along with urban poverty. Exclusion caused by unbalanced growth appears to have been particularly pronounced during the global COVID-19 pandemic. It worsened the living conditions of marginalized urban communities by making their work uncertain and led to severe wage-cuts, thus aggravating income inequality and urban poverty. In the post-pandemic era, tendencies to social, economic and spatial polarization caused by urban poverty and exclusion can be countered by specific efforts, such as the creation of new types of businesses, adaptation to new ways of working and living and improvements on social governance models [22].

The multidimensionality of inclusiveness in urban development needs to be disentangled before it can be used to help cities in shaping policy initiatives or developmental projects [6]. Some authors describe this unpacking as a process. For example, [15] noted that inclusion in the context of construction management is a "process of valuing, respecting and supporting members of an entity" (p. 243) whereas [42] mentioned that it is a "way to increase efficiency in city management and service delivery across urban and peri-urban areas" (p. 277). The deliverable in following this process is meant to be new norms of practice and a change in institutional procedures in city governance. [29] examined the links between inclusion, inequality and place from a "rights-based" perspective and emphasizes human's relationships with the natural environment. In itself, it is apparent that the political, social, economic and environmental dimensions are all main aspects of the inclusive city. Besides, [37] further argued that promoting inclusive and sustainable development should also take into account the cultural dimension of cities based on a participatory process, push factors (subsidies and institutional framework) and societal mobilization.

[9] explain that the inclusive city consists of citizen-centric democratic governance processes, in which governments and other stakeholders need to consider various claims to inclusion put on the table by different groups. They argue that modernity has led to a redefinition of inclusion as an absolute moral imperative that the government and other stakeholders should realize together. Thus, it leads to the need for well-considered and practical tradeoffs based on stakeholder-oriented governance and moral leadership. This point of view also appears in other studies, such as [20].

Similarly, how inclusive urban development as a paradigm shift affects the most vulnerable people and ecological standards and how persistent imbalances in power perpetuate inequality and injustice are described and debated in various academic publications, where it is argued that inclusive cities are to offer a broad range of choices for development, governance and management [20, 28]. They offer a variety of suggestions for inclusive urban development through good governance, networks, instruments and policies largely aimed at the realization of SDGs [28].

Going beyond [40], in this study we explored the most recent interpretations of the inclusive city concept as a supplement and enrichment of previous studies. In terms of urban resilience and regeneration, [12] emphasize that inclusion can primarily be seen as a systemic approach adopted by public and private organizations to help people recover their capabilities, pursue their most important goals and thus pave the road to realizing this ideal. They make inclusive city practical through the notion of recovery and explicitly state that "Inclusive Cities is an initiative to support the creation of Recovery-Oriented Systems of Care at a city level, that starts with but extends beyond substance using populations" [12].

From the perspective of culture, an inclusive city should be open to different subcultures and be the connection point between these different groups. Due to the rubbing of different subcultures against each other, everyone slowly becomes part of a shared society. The design and the activities in the public space and the programming of the buildings ensure that the subcultures enter into contact with, and learn from, each other. Some authors propose that the design of public space plays a role in making inclusive cities possible and offer concrete suggestions. For example, [75] gives the example of more explicitly involving women and girls in the design of public space. Interestingly, [16] expand the function by stating that spatial design is required to foster the inclusive city both from a social and from an environmental standpoint. They state that "the objective of an 'inclusive city' is often related to social issues, which might easily lead to the exclusion of ecological values,the opposite approach may prove equally exclusive. Inclusivity also means creating room for the unexpected … [and] … is inherent to a complete understanding of landscape architecture" [16].

Local governments would do well to strengthen and advocate an inclusive participatory approach in providing access to basic social benefits, such as housing, various infrastructures and public space to make them more accessible to vulnerable groups such as the elderly and the disabled [47, 51, 64]. Identifying active citizenship, empowerment and partnerships between mainstream organizations and people with disabilities are recognized as the most important criteria for social inclusion [55]. Gendered inclusion has also attracted special attention [44, 69, 73]. These authors examine contemporary urban movements as gendered resistance to reclaim space and inclusivity at different scales, and provide a theoretical framework to describe gendered resistance as a means to realize inclusive and sustainable urban spaces [21]. The role of partnerships in creating safe communities for women is of great importance. Such partnerships between women-centered groups, local governments and other relevant partners advance women's interests through closer collaboration [7].

And last but not least, technology-driven urban planning and governance of infrastructures and public services have brought about a dramatic change in the urban development paradigm. When specific social objectives and regulatory frameworks are in focus, the use of data-based tools largely contributes to making cities more inclusive by enhancing spatial inclusiveness and improving the efficiency of urban governance [67]. Besides, it is suggested that innovative modes of action should be adopted and the use of new technologies and digital policies adjusted in such a way that socioeconomic disparities and environmental crises are addressed in unconventional ways [10].

Although authors above differ in their disciplinary perspectives and levels of moral indignation and idealism, we can see a consensus that the inclusive city concept mainly encompasses a social, spatial, environmental, political, economic and a cultural dimension. Below, we offer an overview of which dimension(s) and key terms can be found in the work of which authors (see Table 5).

Table 5 Key terms of each dimension from all authors

	Dimension	Key terms
1	Spatial inclusion	Affordable housing [13, 37, 38, 53, 62, 68], public space [10, 28, 47, 49, 68, 73], transportation and other basic infrastructures [31, 38, 47, 73], spatial justice [47], infrastructure needs [51], service delivery [51], urban spaces [21], disability [34]; gender [26]; gendered resistances [21]; vulnerabilities [26]
2	Social inclusion	Right to the city [24, 38, 47, 49, 73], a sense of security [24, 53, 73]; citizens' rights [24], human rights [47], social justice [9, 13, 47, 62], social equity [72], social participation [38, 47, 55, 68], public services [38, 47, 53], access to information [47], quality of life [38, 55], access to basic services [20]; housing [51]
3	Environmental inclusion	Environmental sustainability [72], solid waste management [38, 49], reduce water loss, make up for lost water and conserve water [49], the natural and reproductive qualities of urban space [9], urban greening [8]; resilience [57]; land clean-up and greening [8]
4	Economic inclusion	Employment [73], inclusive growth [9, 72], shared prosperity [9], diversion of economy [10], green growth [37]
5	Political inclusion	Political participation [47, 55], political empowerment [9, 24, 55, 72], active citizenship [55]; civic participation [11]; stakeholders [11]
6	Cultural inclusion	Diversity [14]; belonging [14]

5 Discussion

The combination of quantitative analysis of academic articles and qualitative analysis of international policy reports and scientific books allows us to extract the following six dimensions of the inclusive city:

Cluster 1, 3 and 7, albeit covering different aspects, all fit in the 1st dimension of *spatial inclusion*. It is often seen as a process of equal access to the essential living environment encompassing land, streets, housing and public infrastructure and facilities for all individuals. Spatial inclusion often depends on the degree to which public space, physically and socially, is open to all. Disabled inclusion and gendered inclusion are two typical facets of spatial inclusion. People with disabilities and women are often highly implicated in the design of the built environment, public transportation and urban form so that they are able to occupy and use urban spaces without fear or discrimination. Paying special attention to each stage of urban planning and enhancing their experience of urban space would give them opportunities shape the urban environment in which they live, which greatly contributes to creating a more accessible, sustainable and inclusive city. For greater universal inclusivity, the focus should be placed on the different needs of the groups. Innovative technologies are playing an incremental yet vital role in addressing the conflict between the rapid expansion of urban land and the provision of adequate space for citizens in an inclusive city.

Second, clusters 1 and 3 from the bibliometric analysis and dimension 2 from the qualitative review primarily cover aspects of *social inclusion*. What social inclusion focuses on is equal development opportunities and attending to social members' needs [5]. Sustainable migration and public participation are two significant characteristics of social inclusion, the former being reflected in the entitlement to decent and affordable accommodation and protection from forced eviction (showing some overlap with spatial inclusion) and the latter denoting the public's concern about social affairs and the level of social acceptance and integration. In addition, all individuals and social groups should have equal access to social resources (e.g., employment, insurance, education, information), and their rights should be protected and secured in situations of vulnerability with diseases, crime, violence, food and accidents.

Third, cluster 3 and 4 match dimension 3 quite well. *Environmental inclusion* implies meeting the needs of current generations for natural sources and environment without compromising the interests of future generations. Meanwhile, it emphasizes close and inseparable relationships between the allocation of resources, environmental pollution and responsibilities [54]. Nowadays, a growing number of local governments have called on the city council, public and private organizations, social communities, employers and the public to work together to promote and facilitate sustained recovery. It shows that local governments give full play to the vital role of the broader public and social organizations in addressing environmental issues (e.g., climate change, air pollution and sewage disposal). It aims to make regeneration visible, to celebrate it and to create a safe environment supportive to urban recovery.

Fourth, cluster 2 from the bibliometric analysis and dimension 1 from the qualitative review neatly fit together in a dimension on *economic inclusion*. Economic inclusion makes it possible for all people, especially the disadvantaged and low-income groups, to share in rising prosperity, i.e. to share in and contribute to gains in welfare and well-being [61]. In fact, in terms of labor market relations and resource allocation, economic inclusion is also considered as a process of eliminating economic inequities caused by rapid urbanization and industrialization along with changing technologies and demand for various skills through a series of policy reforms encompassing equal access to job opportunities, labor market information and reasonable distribution of income. For instance, local governments can step up investment in manufacturing and utilities, encourage migrants to establish new linkages with industries in cities, provide employment opportunities and vocational training for young, women and those in underdeveloped regions, and strengthen their supervision of harmful effects of market activities. More specifically, the informal economy (e.g., street vending) can be seen in a new light, i.e. as a way to promote urban economic regeneration.

Fifth, clusters 3 and 7 match dimension 5 of *political inclusion*, which can be defined as a rational and non-discriminatory citizen-state relationship based on civil and political rights, more precisely a citizen's sense of belonging and identity and his/her empowerment (especially in Western countries). It is thus primarily related to major issues of democratic institutions, human rights, political participation, and

Exploring the Inclusive City: Definitions and Dimensions

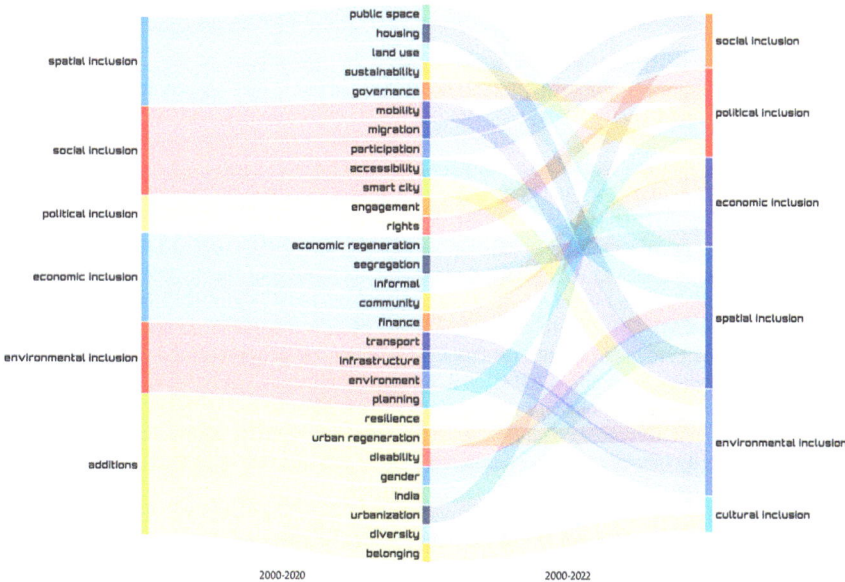

Fig. 6 Changes in the dimensions and connotations of the inclusive city

national identity. It provides channels for effective communication between local governments and citizens and creates a way for citizens to make their claims.

Seventh, cluster 7 can be seen as dimension 6 of *cultural inclusion*. Increasingly urbanized and multicultural existence and its spatial structures and complications require us to reexamine urban inclusion. It can be seen as a new perspective for reading how inclusive city is for various groups with different patterns of values and norms. Cultural heterogeneity and diversity within cities is to be taken into account in the urban policy-making process, so that personal cultural belonging can enable people to garner mutual recognition and respect.

Figure 6 is a graphic display of the dimensions identified above that takes the focal points that have emerged in the last two years into account. It also presents the similarities and differences of dimensions and connotations of inclusive cities between this study and [40]. Nonetheless, overall, we can conclude that the quantitative and qualitative analyses result in rather similar outcomes.

Although the above six dimensions above can be clearly distinguished from each other, they are interwoven and mutually complementary: there are synergistic effects between them in enhancing inclusiveness of the city as a whole. Creating an inclusive city can be seen as a complex practice both intellectually and politically, because it is comprised of different dimensions for which a form of coordination is to be found in governance, policy making and management to accommodate various partly divergent stakeholder interests.

References

1. Adebayo P, Ndinda C, Ndhlovu T (2022) South African cities, housing precarity and women's inclusion during COVID-19. Agenda 36(12):16–28. https://doi.org/10.1080/10130950.2022.2057027
2. Affleck A, Mellor M (2006) Community development finance: a neo-market solution to social exclusion? J Soc Policy 35:303–319. https://doi.org/10.1017/S0047279405009542
3. Agost-Felip R, Ruá MJ, Kouidmi F (2021) An inclusive model for assessing age-friendly urban environments in vulnerable areas. Sustain 13(15):8352. https://doi.org/10.3390/su13158352
4. Aksnes DW, Sivertsen G (2019) A criteria-based assessment of the coverage of scopus and web of science. J Data Inf Sci 4(1):1–21. https://doi.org/10.2478/jdis-2019-0001
5. Albuquerque CMP (2017) Cities really smart and inclusive: possibilities and limits for social inclusion and participation. In: Handbook of research on entrepreneurial development and innovation within smart cities. IGI Global Scientific Publishing, pp 229–247
6. Alsayel A, de Jong M, Fransen J (2022) Can creative cities be inclusive too? How do Dubai, Amsterdam and Toronto navigate the tensions between creativity and inclusiveness in their adoption of city brands and policy initiatives? Cities 128:103786. https://doi.org/10.1016/j.cities.2022.103786
7. Andrew C, Legacy C (2013) The role of partnerships in creating inclusive cities. In: Building inclusive cities: women's safety and the right to the City, 1st edn. Routledge, pp 90–102
8. Anguelovski I, Connolly JJ, Cole H, et al (2022) Green gentrification in European and North American cities. Nat Commun 13(1):3816. https://doi.org/10.1038/s41467-022-31572-1
9. Anttiroiko A-V, De Jong M (2020) The inclusive city: the theory and practice of creating urban prosperity for all. Springer Nature
10. Attia S, Shafik Z, Ibrahim A (eds) (2019) New cities and community extensions in Egypt and the Middle East: visions and challenges. Springer, New York. https://doi.org/10.1007/978-3-319-77875-4
11. Banerjee N (2019) Community-driven development as a mechanism for realizing global development goals: the National Solidarity Programme and Citizens' Charter Afghanistan Program. In: Better spending for localizing global sustainable development goals. Routledge, pp 137–150
12. Best D, Colman C (2019) Let's celebrate recovery. Inclusive Cities working together to support social cohesion. Addict Res Theory 27(1):55–64. https://doi.org/10.1080/16066359.2018.1520223
13. Bharne V, Khandekar S (eds) (2019) Affordable housing: inclusive cities. ORO Editions, New York
14. Blanchet-Cohen N, Torres J, Grégoire-Labrecque G (2020) Youth and their multiple relationships with the city: experiences of exclusion and belonging in Montréal. In: Rethinking young people's lives through space and place. Emerald Publishing Limited, pp 85–103
15. Blay K (2018) The impact of inclusiveness on resilience in Temporary Multidisciplinary Organizations (TMO). Construction research congress 2018, pp 243–252
16. Bobbink I, de Wit S (2021) Landscape architectural perspectives as agent for generous design. Res Urban Ser 6:129–149. https://doi.org/10.7480/rius.6.97
17. Briceño-León R (2022) Feral cities and the normative dimension of violence: Caracas and the Latin American city. Urban violence, resilience and security: governance responses in the Global South. Edward Elgar Publishing Ltd., Laboratorio de Ciencias Sociales (LACSO), Universidad Central de Venezuela, Universidade Federal do Ceará, Brazil, pp 101–119
18. Chang J-I, Choi J, An H, Chung H-Y (2022) Gendering the smart city: a case study of Sejong City, Korea. Cities 120:103422. https://doi.org/10.1016/j.cities.2021.103422
19. Chatterjee A (2021) Contemporary urban missions and reflecting reality in deprivation of civil areas in Indian Cantonments-a pragmatic view. J Settlements Spat Plan 12(2):71–81. https://doi.org/10.24193/JSSP.2021.2.01
20. Dahiya B, Das A (2020) New urban agenda in Asia-Pacific: governance for sustainable and inclusive cities. In: Dahiya B, Das A (eds) New urban agenda in Asia-Pacific. Advances in 21st

century human settlements. Springer, Singapore. https://doi.org/10.1007/978-981-13-6709-0_1
21. Datta A (2021) Gender, urban spaces and gendered resistances: towards inclusive and fear free cities in India. In: Jaglan MS, Rajeshwari (eds) Reflections on 21st century human habitats in India. Advances in 21st century human settlements. Springer, Singapore. https://doi.org/10.1007/978-981-16-3100-9_13
22. Dávila JD (2020) Urban mobility and social equity in Latin American cities: evidence, concepts and methods for more inclusive cities. In: Oviedo D, Duarte, NV and Pinto AMA (eds) (Transport and Sustainability), Emerald Publishing Limited, Leeds, pp 235–237. https://doi.org/10.1108/S2044-994120200000012017
23. van Eck NJ, Waltman L (2014) Visualizing bibliometric networks. In: Ding Y, Rousseau R, Wolfram D (eds) Measuring scholarly impact. Springer, Cham, pp 285–320. https://doi.org/10.1007/978-3-319-10377-8_13
24. Espino NA (2015) Building the inclusive city: theory and practice for confronting urban segregation (1st ed.). Routledge, London and New York
25. Eyita-Okon E (2022) Urbanization and human security in post-colonial Africa. Front Sustain Cities 4:917764. https://doi.org/10.3389/frsc.2022.917764
26. Faret L (2021) Has Mexico city truly become a ciudad hospitalaria? Insights from the experience of central American migrants. In: Faret L, Sanders H (eds) Migrant protection and the city in the Americas. Politics of citizenship and igration. Palgrave Macmillan, Cham. https://doi.org/10.1007/978-3-030-74369-7_8
27. Fredericks J, Hespanhol L, Parker C et al (2018) Blending pop-up urbanism and participatory technologies: challenges and opportunities for inclusive city making. City Cult Soc 12:44–53. https://doi.org/10.1016/j.ccs.2017.06.005
28. Gupta J, Pfeffer K, Verrest H, Ros-Tonen M (eds) (2015) Geographies of urban governance: advanced theories, methods and practices. Springer. https://doi.org/10.1007/978-3-319-21272-2
29. Hambleton R (2014) Leading the inclusive city: place-based innovation for a bounded planet. Policy Press, Bristol
30. Henderson-Wilson C, Andrews F, Wilson E, et al (2022) Global Benchmarking of Accessible and Inclusive Cities. J Soc Incl 13(1):42–65
31. Herrle P, Walther U-J (eds) (2005) Socially inclusive cities: emerging concepts and practice. Transaction Publishers, New Jersey
32. Huang X, Song J, Wang C, Chan PW (2022) Realistic representation of city street-level human thermal stress via a new urban climate-human coupling system. Renew Sustain Energy Rev 169:112919. https://doi.org/10.1016/j.rser.2022.112919
33. Jetoo S (2019) Stakeholder engagement for inclusive climate governance: the case of the City of Turku. Sustainability 11:6080. https://doi.org/10.3390/su11216080
34. Kamuzhanje J (2021) Urbanisation, inclusive cities and the plight of the people with disability. In: Magidimisha-Chipungu HH, Chipungu L (eds) Urban inclusivity in Southern Africa. The urban book S=series. Springer, Cham. https://doi.org/10.1007/978-3-030-81511-0_7
35. Keen M, Ride A (2019) Trading places: Inclusive cities and market vending in the Pacific Islands. Asia Pac Viewp 60(3):239–251. https://doi.org/10.1111/apv.12227
36. Kharel S (2017) Rural womens' access to community finance. Nepal J Dev Rural Stud 14(1–2):112–123. https://doi.org/10.3126/njdrs.v14i1-2.19654
37. Kundu D, Sietchiping R, Kinyanjui M (eds) (2020) Developing national urban policies: ways forward to green and smart cities. Springer, Singapore. https://doi.org/10.1007/978-981-15-3738-7
38. Laquian AA, Tewari V, Hanley LM (eds) (2007) The inclusive city: infrastructure and public services for the urban poor in Asia. Woodrow Wilson Center Press/Johns Hopkins University Press, Baltimore
39. Li W, Xu J, Zheng M (2018) Green governance: new perspective from open innovation. Sustain 10:3845. https://doi.org/10.3390/su10113845

40. Liang D, De Jong M, Schraven D, Wang L (2022) Mapping key features and dimensions of the inclusive city: a systematic bibliometric analysis and literature study. Int J Sustain Dev World Ecol 29(1):60–79. https://doi.org/10.1080/13504509.2021.1911873
41. Marta B, Giulia D (2020) Addressing social sustainability in urban regeneration processes. An application of the social multi-criteria evaluation. Sustain 12(18):7579. https://doi.org/10.3390/su12187579
42. McCarney P (2010) Conclusions: governance challenges in Urban and Peri-urban Areas BT - Peri-urban water and sanitation services: policy, planning and method. In: Kurian M, McCarney P (eds) Peri-urban water and sanitation services. Springer, Netherlands, Dordrecht, pp 277–297
43. Moghayedi A, Richter I, Owoade FM et al (2022) Effects of urban smart farming on local economy and food production in urban areas in African cities. Sustain 14(17):10836. https://doi.org/10.3390/su141710836
44. Ndinda C, Adebayo P (2021) Human settlement policies and women's access to the city: implications for inclusive cities. In: Magidimisha-Chipungu HH, Chipungu L (eds) Urban inclusivity in Southern Africa. The urban ook Sesries. Springer, Cham. https://doi.org/10.1007/978-3-030-81511-0_15
45. Nyamai DN, Schramm S (2022) Accessibility, mobility, and spatial justice in Nairobi, Kenya. J Urban Aff 45(3):367–389. https://doi.org/10.1080/07352166.2022.2071284
46. OECD (2016) Making cities work for all: data and actions for inclusive growth. OECD Publishing, Paris
47. Pineda VS (2020) Building the inclusive city: governance, access, and the urban transformation of Dubai. Palgrave Pivot
48. Pokharel S, McDonald K, Arup SA (2020) Child-centred urban resilience framework: A tool for inclusive city planning. Aust J Emerg Manag 35(2):7–8
49. Pokhrel N (ed) (2019) Transforming Kolkata: a partnership for a more sustainable, inclusive, and resilient city. Asian Development Bank, India
50. Poku-Boansi M (2021) Multi-stakeholder involvement in urban land use planning in the Ejisu Municipality, Ghana: an application of the social complexities' theory. Land Use Policy 103:105315
51. Popoola AA, et al (2021) The language of struggle and radical activism as an inclusive city tool among the neglected urban poor of South Africa. In: Magidimisha-Chipungu HH, Chipungu L (eds) Urban inclusivity in Southern Africa. The urban book series. Springer, Cham. https://doi.org/10.1007/978-3-030-81511-0_19
52. Salahub JE, Gottsbacher M, de Boer J (eds) (2018) Social theories of urban violence in the global south: towards safe and inclusive cities. Routledge, London and New York
53. Salahub JE, Gottsbacher M, De Boer J, Zaaroura MD (eds) (2019) Reducing urban violence in the global south: towards safe and inclusive cities. Routledge, London and New York
54. Sands P, Peel J (2018) Principles of international environmental law. Cambridge University Press, Cambridge MA
55. Schippers A, Van Heumen L (2014) The inclusive city through the lens of quality of life. In: Quality of life and intellectual disability: knowledge application to other social and educational challenges. Nova Science Publishers, Inc., New York
56. Schraven D, Joss S, De Jong M (2021) Past, present, future: engagement with sustainable urban development through 35 city labels in the scientific literature 1990–2019. J Cle Prod 292:125924. https://doi.org/10.1016/j.jclepro.2021.125924
57. Sharma S, Batra N (2019) Comparative study of single linkage, complete linkage, and ward method of agglomerative clustering. In: 2019 international conference on machine learning, big data, cloud and parallel computing: trends, prespectives and prospects (COMITCon), pp 568–573
58. Shobeiri S (2021) Inclusiveness in street network of city centre—case studies: 15-Khordad, Berlan and Sepah-Salar pedestrian-based axes in central Tehran. Environ Ecol Res 9(1):1–29. https://doi.org/10.13189/eer.2021.090101
59. Singru RN, Lindfield MR (2017) Enabling inclusive cities: tool kit for inclusive urban development. Asian Development Bank, Mandaluyong

60. Sirkeci I, Murat Yüceşahin M (2020) Coronavirus and migration: analysis of human mobility and the spread of COVID-19. Migr Lett 17(2):379–398
61. de Souza BX, Pendall R, Rubin V (2015) Inclusive economic growth in America's cities: what's the playbook and the score? Social science electronic publishing. World bank policy, Research Working Paper No. 7322. Washington, DC
62. Steinberg F, Lindfield MR (2011) Inclusive cities. Mandaluyong
63. Treija S, Bratuškins U, Koroļova A (2022) University-community engagement: formation of new collaboration patterns in participatory budgeting process. Archit Urban Plan 18(1):156–165. https://doi.org/10.2478/aup-2022-0016
64. Tucker R, Kelly D, Johnson L, De Jong U, Watchorn V (2022) Housing at the fulcrum: a systems approach to uncovering built environment obstacles to city scale accessibility and inclusion. J Hous Built Environ 37(3):1179–1197. https://doi.org/10.1007/s10901-021-09881-6
65. UN-Habitat III (2015) Habitat III issue paper 1-inclusive cities. New York
66. UNCHS (Habitat) (2001) The state of the world's cities, 2001. UN-HABITAT, New York
67. Uteng TP, Christensen HR, Levin L (2020) Gendering smart mobilities. Taylor and Francis, Institute of Transport Economics, Oslo, Norway
68. Venkateswar S, Bandyopadhyay S (eds) (2016) Globalisation and the challenges of development in contemporary India (dynamics of Asian development). Springer, New York
69. Viswanath K (2013) Gender inclusive cities programme: implementing change for women's safety. In: Whitzman C, Legacy C, Andrew C, et al. (eds) Building inclusive cities: women's safety and the right to the city. Routledge, New York
70. Wahba SN (2022) Can cities bounce back better from COVID-19? Reflections from emerging post-pandemic recovery plans and trade-offs. Environ Urban 34(2):481–496
71. Wang X, Liu Z (2022) Neighborhood environments and inclusive cities: an empirical study of local residents' attitudes toward migrant social integration in Beijing, China. Landsc Urban Plan 226:104495
72. Westendorff D (ed) (2004) From unsustainable to inclusive cities. UNRISD, Geneva
73. Whitzman C, Legacy C, Andrew C et al (2013) Building inclusive cities: Women's safety and the right to the city. Routledge, New York
74. World Bank Group (2015) World-inclusive cities approach paper. NW Washington, DC
75. Yang H, Berry J, Kalms N (2022) Perceptions of safety in cities after dark. In: Lighting design in shared public spaces, pp 83–103

Open Access This chapter is licensed under the terms of the Creative Commons Attribution 4.0 International License (http://creativecommons.org/licenses/by/4.0/), which permits use, sharing, adaptation, distribution and reproduction in any medium or format, as long as you give appropriate credit to the original author(s) and the source, provide a link to the Creative Commons license and indicate if changes were made.

The images or other third party material in this chapter are included in the chapter's Creative Commons license, unless indicated otherwise in a credit line to the material. If material is not included in the chapter's Creative Commons license and your intended use is not permitted by statutory regulation or exceeds the permitted use, you will need to obtain permission directly from the copyright holder.

A Conceptual Framework for Inclusive Circular Urban Waste Management Systems

Daan Schraven, Filippos K. Zisopoulos, Liang Dong, and Martin de Jong

Abstract Rapid urbanization in combination with unsustainable production and consumption patterns leads to the generation of substantial amounts of urban waste. The circular economy promises to bring solutions both with top-down and with bottom-up activities. The former relate to the implementation of policies which are based on the waste hierarchy principles by local governments, whereas the latter are about the adoption of circular business models by urban stakeholders. However, a circular economy does not automatically endow cities with inclusion or resilience against shocks. Consequently, any decision which relates to such a transition is not trivial. This chapter presents an integrative framework to assist urban decision makers in considering inclusion and circularity simultaneously when developing urban waste management systems where urban regeneration has a central role. The framework places explicit attention on improving the accessibility of social groups to various forms of capital and stimulating the development of local economies through improved circulation of resources and information within the urban fabric.

Keywords Inclusion · Urban waste management system · Circularity

---

D. Schraven (✉) · F. K. Zisopoulos
Faculty of Architecture in the Built Environment, Management in the Built Environment (MBE), Section of Real Estate Management, Delft University of Technology, Delft, The Netherlands
e-mail: d.f.j.schraven@tudelft.nl

F. K. Zisopoulos · M. de Jong
Rotterdam School of Management & Erasmus School of Law, Erasmus University Rotterdam, Rotterdam, The Netherlands

L. Dong
Department of Public and International Affairs (PIA), and, School of Energy and Environment (SEE), City University of Hong Kong, Hong Kong, China

M. de Jong
Smart City Institute, HEC Liege, Liege, Belgium

Institute of Global Public Policy, Fudan University, Shanghai, China

1 Introduction

Realizing inclusive circular cities aims to eliminate linear waste flows in which the development and management of urban waste management systems (UWMS) as resource collection and distribution networks play an important role by involving various societal stakeholders, including vulnerable ones. Since the establishment of waste sorting facilities along with waste-to-energy plants represents a long-term investment and any transition to a different type of waste management system is hard and costly, potentially harmful lock-ins due to the construction of a suboptimal waste management infrastructure should be avoided. This is particularly important in countries where substantial power asymmetries between urban actors might exist, where the design process may get politicized, and where consultation with local stakeholders can be easily overlooked. Consequently, UWMSs need to be developed along *circularity principles* whereby all stakeholders act responsibly and adhere to *inclusion principles*, such that vulnerable but relevant groups can also do their say; that task is not easy to achieve.

This chapter aims to establish a framework to study the development of UWMSs in which both circularity and inclusion principles are respected. Section 2 introduces a design procedure, called the Double Diamond, and subsequently applies it to (a) discover the implications of the key circularity and inclusion principles in an urban context, (b) define the resulting design requirements, (c) develop a rationale for key input and output elements to assess the circularity and inclusion of UWMSs, and (d) deliver a final framework based on these suggestions. Finally, Sect. 4 discusses the implications of the framework and highlights that circularity and inclusion both provide important guidelines for the development of UWMSs which can boost the sustainability performance of cities. In Sect. 5, conclusions are drawn.

2 Applying the Double Diamond to Produce a Conceptual Framework

2.1 What is the Double Diamond Method?

The **Double Diamond (DD) method** is a design-based research method known for its diverging and converging phases in which a problem at hand is explored and a solution developed (British Design Council 2007). The consideration of diverging and converging features of the DD method fit well with the purposes of this chapter because the challenge which needs to be solved has three tiers. The first tier is making a conceptual linkage between circularity and inclusion across different levels (city, its UWMS, and actors and physical facilities). The second tier is the contextual embedding for a clear translation of the circularity and inclusion principles into integrated UWMSs. The third tier is the tailoring nature of the chosen design approach: it needs to be applicable to different urban contexts and adaptable to various requirements.

These three tiers benefit from an open consultation step followed by strict evaluation steps.

The research design based on the DD method follows four consecutive phases. The **discover phase** is a first diverging step where the dimensions of an existing problem are investigated. In the **define phase** the emerging opportunities to deal with the problem are analyzed and synthesized into a development plan. This plan is then used as input for the **develop phase** to initiate, elaborate and evaluate the ideas for their usefulness. Finally, the **deliver phase** guides the implementation of the ideas by iteratively building and testing the chosen design. The next section explains how these research steps evolve.

2.2 DD as a Research Process in Four Phases

To facilitate going through the empirical steps, this study was conducted in three workshops with researchers of the Inclusive Wise Waste Cities (IWWC) project by following the research setup shown in Fig. 1. Each phase in this research setup included methodological choices which require justification. Below, the main points are described and tagged with brackets [...] corresponding with the numbers of Fig. 1.

Discover phase: Here, an actual problem analysis was set prior to the design-objective (DO) and guided by a combination of secondary and primary research. It was important to set aside any prior assumptions about the problem so that the relevant circularity and inclusion requirements for an UWMS could be identified (e.g., via second-hand data derived from other studies). General insights regarding circularity and inclusion were identified from literature study and presented above in Chapters 2 and 3. Thereafter, specific characteristics of UWMSs were identified during a brainstorming workshop among six IWWC researchers.

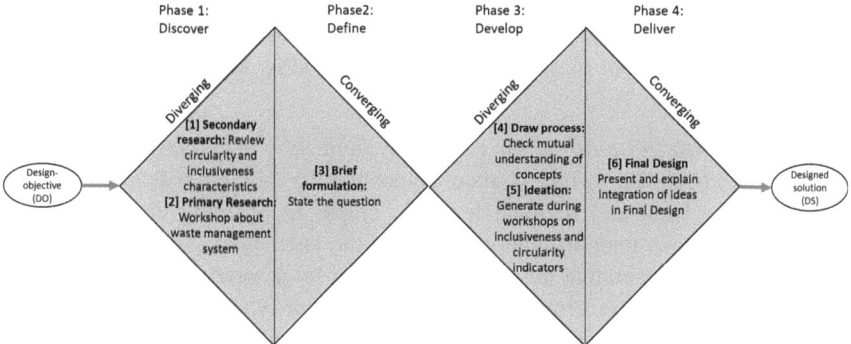

Fig. 1 Setup following the double diamond design-based research method

Define phase: During this phase the output of the discovery phase was analyzed and synthesized in pursuance of providing a clear brief. Moreover, expectations and conditions were formulated as part of the design brief.

Develop phase: This phase generated tangible ideas to lay the foundation for a design with input from the preceding phase. It required a clear target for further elaborating the ideas. We focused on the ways in which circularity and inclusion can be measured and evaluated in the ways UWMSs are operated. A fallback was organized by means of two additional workshops with members of the development team to create a common understanding around circularity and inclusion. Ideas were then generated covering the input from the define phase to the best possible degree and to assure meaningful incorporation of both circularity and inclusion related requirements in a preliminary framework.

Deliver phase: At this phase no designed elements were permanently established. Rather, topics remained subject to possible changes which allowed for further refinement before a final conceptual framework could be produced. This step helped to secure the appropriate and coherent positioning of various elements and relations within that final framework.

3 Towards a Conceptual Framework for Inclusive Circular UWMSs

In this section, we apply the four phases of the DD method to deliver a framework for inclusive circular Urban Waste Management Systems.

3.1 Discover

A broad search in the academic literature highlighted several challenges associated with the development of inclusive and circular UWMSs. More specifically, a city needs:

- *to adopt strategies tailored for circularity both from a top-down and bottom-up perspective.* The former is about implementing circular policies and directives at the local and regional levels to create an enabling environment for networked ecosystems to emerge, whereas the latter refers to the adoption of circular and regenerative business models [31] by a variety of urban actors to facilitate the effective cross-circulation of different types of resources such as materials, energy, money, and information [e.g., narrowing, slowing, intensifying, and dematerializing resources [5, 7, 24]].
- *to be able to bring on-board a plurality of actors both from the public and the private sectors* including stakeholders who are not normally considered as part of the industrial cohort (e.g., NGOs, banks, and citizens amongst others) to jointly

identify relevant opportunities and challenges in the local context. Here, transition brokers can play an instrumental role for the practical implementation of CE initiatives because they can facilitate the alignment of interests between regime and niche actors, and bolster their collaborative efforts [15].

- *to be able to measure and track circularity and foster inclusion across geographical scales.* Circularity indicators are often rather technical in orientation to address different issues at the micro-, meso-, and macro-levels. Multiple tools and frameworks are constantly being developed to meet these needs. For example, [20] proposed a systematic framework to help decision makers in selecting appropriate methods for assessing the implementation of CE strategies [4] listed thirty-five waste-related indicators and twenty-four circular economy indicators. Ten out of those indicators were categorized by the authors as dashboard indicators since they are meant to provide a summarized perspective on the performance of an UWMS within a given area.

- *to ensure that marginalized groups are not excluded or left behind.* A city can considerably improve the efficiency of its UWMS by empowering informal recyclers who with their tacit knowledge contribute substantially to the collection and recycling activities for various types of solid waste streams. In this way, they create value for their cities as key economic and environmental actors [17], they help in building resilience, and they can even help in tackling several of the UN Sustainable Development Goals [27]. Their contribution varies considerably across many low- and middle-income countries but it is difficult to quantify and generalize waste collection and recycling rates due to their invisibility in the formal economy and their stigmatization, discrimination, and otherwise, exclusion from the society at large [17, 28, 37, 56, 60]. Therefore, the inclusion of informal recyclers is crucial for the transition towards a CE but only when issues which typically include (but are not limited to) the recognition of their professions and enterprises, the legalization of opportunities, and their integration in reliable, fair, and participatory positions in discussions [46], are structurally addressed.

- *to deploy infrastructure which informal recyclers can access easily* to participate actively, and where their contribution can be sustained as a regenerative capability [33]. In this way, they can offer their selective waste collection and resource recovery services to the city while at the same time improving social cohesion [28]. Secondary markets can dictate the emergence or decline of the informal sector in a city, which are of importance as they can be more effective than the formal ways of handling waste [3]. However, the inclusion of informal recyclers in UWMSs should not merely serve the expansion of the capabilities of recycling facilities but also actively support sustainable community initiatives [54]. To offer better and decent jobs related to resource management, local authorities will need to work together with social organizations [26]. It makes sense that the design process for UWMSs only takes place before the implementation of any CE policies to facilitate the identification of potential impact and shape effective cooperation while avoiding any risks of worsening livelihoods [47].

With these key insights from the academic literature in mind, a team of six researchers held a workshop with the aim to answer three questions, emphasizing on the rationale of the framework and focusing on potentially missing elements. Those questions concerned the necessity of the framework, the required know-how, and its future application. The main points highlighted during the workshop were that such a framework should:

1. *clearly demarcate the boundaries of the UWMS.* A metaphor was made to specify what was meant by boundaries: these were likened to the human body where the skin serves as its physical boundary and its organs and bones form the infrastructure of the larger system. Here, it is important to identify which stakeholders and physical elements are part of the UWMS [e.g., by following the economy-as-an-organism analogy propositions of [36]], and to clarify which system elements are to be analyzed and which not.
2. *enable the examination of possibilities to optimize effectiveness and handling emerging trade-offs within the system boundaries of the UWMS.* Following the analogy of the human body, one would expect the framework to target the optimization of the function of certain organs which might come at the cost of sub-optimizing other parts elsewhere in the organism. In this way, additional knowledge could be generated about the systemic limits in terms of circular performance, inclusive benefits, and emerging contradictions in the function of supporting infrastructure.
3. *allow for the consideration of a plethora of metrics and indicators to capture both dimensions of inclusiveness and circularity* to the best possible degree.
4. *allow for the identification of all relevant stakeholders in the UWMS along with all relevant stocks and flows.* With this knowledge the basic architecture of data requirements can be set up. To this end, methods from socio-metabolic research (e.g., [29, 49]) such as material and energy flow analyses are invaluable.
5. *allow for the examination of the regenerative potential of the UWMS by capturing relevant network properties such as its resilience, robustness, and circulation of resources.* Such an approach has been demonstrated in the assessment of the regenerative potential of the socio-economic metabolism of a small island [65].

3.2 Define

The design brief is formulated in the form of a "how might we...?" (HMW) question, addressing the expectations and expertise needed to answer that type of question:

1. *How might we establish a framework which, if used, can make the UWMS of a city capable of fulfilling the needs to:/*
   1. bring multiple actors on board when designing an inclusive and circular UWMS;
   2. ensure that all stakeholders are empowered to contribute to urban circularity to maximize the benefits from both established and more vulnerable groups;

3. measure circularity across scales;
4. establish supporting infrastructure to realize the points made above.

2. *In order to fulfil the following tasks:*
   1. define system boundaries;
   2. examine how goal achievement can be optimized and trade-offs made in well-considered ways;
   3. identify relevant metrics & indicators;
   4. identify relevant stocks, flows & actors; and
   5. evaluate the regenerative potential of the UWMS.

3. *What are the expectations and expertise needed for an answer?*

   The framework should:

1. invite research based on a mixed methods approach whereby both quantitative (e.g. surveys) and qualitative methodologies (semi-structured interviews; and participatory methods) can be used to capture relevant indicators and aspects of the system which otherwise is not quantifiable (e.g., behaviors, norms, and values).
2. allow for the development of tailor-made indicators to capture particular lifecycle stages of urban projects or plans relevant to the development of the UWMS.
3. account for exclusion aspects as negative side-effects. This can be important knowledge for setting system boundaries. For example, if a certain stakeholder is not supported in participating within an UWMS, its modelled interactions in relevant stocks and flows also ceases to exist. Choices as to how to deal with the removed stakeholder or with what remains of the system after its removal should be made consciously.
4. function as a canvas to facilitate consensus on its main purpose and target audience. It could be evaluative (i.e., judging an existing UWMS with certain criteria), comparative (i.e., comparing various cases to learn from differences and similarities between different UWMSs), or design-based (i.e., a framework which helps to develop an UWMS).
5. allow for a plurality of theories and concepts to fit in (e.g., on topics around the circular society, inclusive circular economy, inclusive capitalism, de-growth, post-growth, sustainable circular economy, strong or weak sustainability, and regenerative economics amongst others).

3.3 Develop

In this section, the design brief is taken as the starting point because it combines the identified challenges at the city level with the functions that the designed framework must meet (see Table 1). First, the input from stakeholders serves as a scoping exercise to identify the system boundaries of the UWMS. Second, the goals for the UWMS in terms of circularity and inclusion are identified. This way, the potential pitfalls which might arise from the designed UWMS can be conceptualized. Third, the relevant indicators for circularity and inclusion representative of the goals can be listed. Fourth, the desired optimization targets for the UWMS and any potentially emerging trade-offs are clearly stated. Fifth, the identified indicators will determine the data requirements and the data collection plan. This involves the collection of raw data as sources or evidence for circularity (e.g., material stocks and flows) and of inclusion or exclusion of relevant actors [e.g. social, political, environmental, economic, spatial, and cultural inclusion as identified in Chapter 3; [1]]. Finally, this process will allow for the first steps towards evaluating the regenerative potential of the UWMS, to be made.

Table 1 can be conceptualized as a network (Fig. 2), which is helpful to understand the subsequent actions below. For example, *"A: Gathering and aligning views of multiple stakeholders"* can be interpreted as a first step whereas *"6: evaluation"* is the final step. In between, there are other operationalization steps which involve the identification of stocks and flows, the identification of metrics and indicators, the testing of possibilities of a solution space for the framework, and the examination of the optimization potential and trade-offs.

Table 1 can then be conceptualized as a sequence of steps (Fig. 3) to cover clear themes. The first step is the engagement of multiple key stakeholders and align their views on common goals. Thereafter, the design of the UWMS can be planned by specifying the scope in terms of its system boundaries and the quantitative model requirements in terms of indicators and data. These steps serve as a preparation for designing and measuring the performance of an UWMS.

Once a preliminary version of the UWMS design is proposed, in the next steps its performance relative to the desired optimization goal and the potential emergence of trade-offs can be assessed. Both the model examination and the real-world engagements are also conceptualized to allow for the evaluation of the regenerative aspects of the UWMS. The evaluation (step 6) is useful for an assessment of the overall performance of the system and can serve as a feedback loop should a need to *"recalibrate"* it arise (steps 7a and 7b). This overall process takes the form of a Plan-Do-Check-Act cycle.

During this cycle, it is important to first identify the *"hardware"* and *"software"* elements constituting the UWMS. The former consists of waste handling and treatment facilities including waste collection, upcycling sites (green points), recycling and reprocessing factories, waste-to-energy plants, and landfills. The latter refers to intangible aspects which guide the process of waste management, such as waste management policies, standards on soil protection, secondary materials, and air quality, promotion activities for desired citizen behavior (e.g., dry separation at the source), the use of relevant software, and extended producer responsibility schemes. Software elements are important as they can largely influence participation rates in the UWMS.

Table 1 Challenges and required functions for developing the framework

No	Necessity for framework Description of challenges (summary of table 2)	Define system boundaries	Examine optimization options	Examine potential trade-offs	Identify relevant metrics & indicators	Identify relevant stocks, flows & actors	Evaluate regenerative potential
1	gathering and aligning the views of multiple stakeholders	X					
2	achieving circularity and enjoying inclusiveness benefits for all stakeholders			X			
3	measuring and quantifying circularity across scales		X		X	X	
4	empowering and facilitating the participation of marginalized groups		X		X	X	
5	establishing supporting infrastructure to prove and improve the capability of marginalized groups		X	X		X	X

3.3.1 Exploring Indicators for Circularity

Indicators for circularity at four different levels

During the workshop, it was acknowledged that circularity can be quantified at various levels, i.e. at the *nano level* (referring to products), *micro level* (company or local level), *meso level* (business ecosystems or regions) and *macro level* (referring to the system-of-systems, or national or international level) [20].

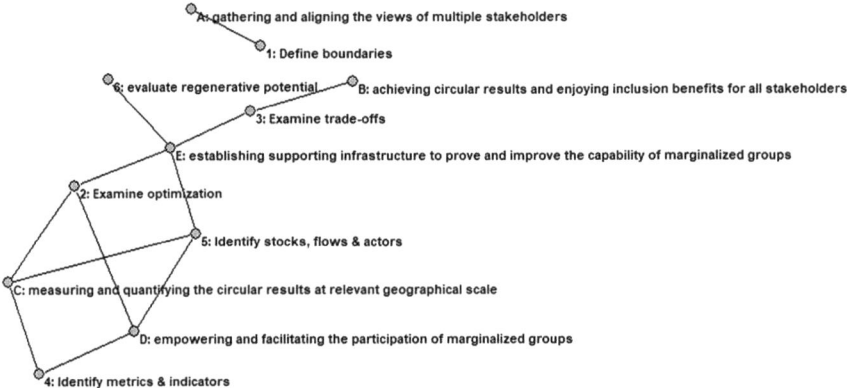

Fig. 2 Connections between description of challenges and necessity for framework

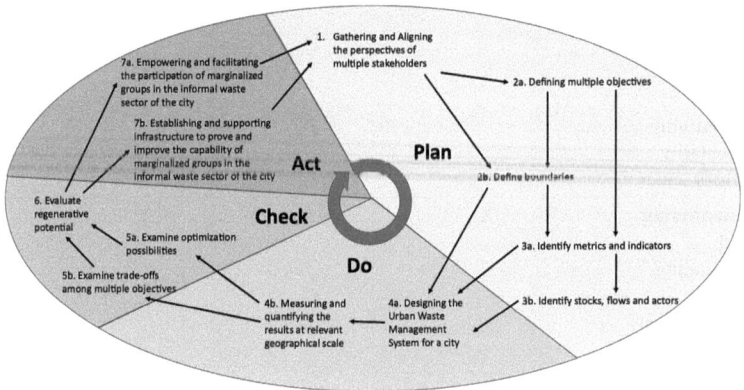

Fig. 3 Methodology for developing inclusive and circular UWMSs

The nano level aims to address the circularity of products: measurements should capture the implementation of closing, slowing, narrowing, and dematerializing loops, all efforts can help track possible reductions in material use.

At the micro level, Life Cycle Assessment (LCA) can be useful for quantifying the social, environmental, and economic impact of various products (e.g., raw material extraction, production, usage of product, and end-of-life management). Moreover, the level of 'demountability' of assets, products, or resources that a company or a local actor works with could be used as a proxy of the agency of an actor to quickly divest certain activities.

At the meso level, material flow analysis can be used to estimate the quantities of materials flowing from one destination to the next, thereby separating their quality in terms of sustainable or circular origin (e.g., with recycled input, or as output planned for reuse). Through industrial symbiosis, eco-innovations can improve mutually beneficial transactions between stakeholders [34]. For individual actors in the emerging ecosystem, two types of responsibility were identified: corporate social responsibility refers to the stakeholder-oriented behavior of corporations to be socially responsible and improve stakeholders' welfare (Liang and Renneboog 2017). Useful indicators for this concept include the provision of employee benefits or environmentally friendly investments. Extended producer responsibility schemes make producers accountable for the environmental impacts throughout the life cycle of products (Gupt and Sahay 2015). There, a potential indicator could be the number of actors in an ecosystem vowing for such responsibility, and the number of visible activities that can be witnessed to curb expired products from becoming waste.

Finally, at the macro level, indicators such as the circular material rate (CMR) of the European Union [22] tend to be used to monitor progress in the transition to a circular economy.

Capturing the circularity for UWMS

Circularity indicators can be categorized in terms of context and content.

Context indicators describe the system from afar, at various scales. At the largest scale, this involves the LCA of the entire resource flow from cradle to cradle/grave, irrespective of system boundaries. Content indicators target the flows up close, both outside and inside the UWMS.

A few indicators address actors in UWMSs. The number of corporate declarations in annual reports is an example of an indicator for capturing aspects of corporate social responsibility. The extent of the materials life cycle that producers haul back beyond the sales (e.g., amount of material that actors have returned into their possession after previously sold products are discarded by their customers) is an example of an indicator which can capture aspects of extended producer responsibility schemes.

3.3.2 Exploring Indicators for Inclusion

The discussion on what inclusion really is gave rise to two different points of view. On the one hand, it was thought to refer mainly to 'everyone benefiting equally of something' and to 'treating people equally without exception'. On the other hand, inclusion was described in terms of participation. This distinction helped to organize the answers to the second question.

The Gini-coefficient is a statistical measure that expresses the inequality between values (like income, wealth or consumption) along a frequency distribution for countries or certain social groups [61]. The Palma ratio expresses the ratio of the share of all income that is received by the 10% richest people divided by the share of income received by the 40% poorest people [39]. The Theil index is used to measure economic inequality and has its origin as an entropy-based metric [13]. The Inclusive Wealth Index differs from other metrics in that it does not focus on Gross Domestic Product but expresses values which measure whether a country develops in a way which allows future generations to meet their own needs with the accumulated wealth (UNEP 2018). This indicator comes close to how the original sustainability definition by Brandt (1988) was intended as passing on the earth to future generations. The Human Development Index (HDI) was developed by the UNDP to emphasize that people and their capabilities should be the ultimate criteria for assessing the development of a country (UNDP 2023). It combines all kinds of social aspects of a country's people in a composite index, like life expectancy, education, and per capita income. This metric has been updated and adjusted various times. The latest time was in 2010, when it was introduced as the inequality-adjusted HDI, or IHDI. This new version corrected for differences in inequalities in a country meant to reflect the actual level of human development, whereas the older version was only focusing on the potential that a country could develop if there were no inequality (UNDP 2023). The Multidimensional Inclusiveness Index (MDI) was introduced by Dorffel and Schuhmann [18]; it is composed of two sub-indices, one on development equity and development achievements. These contain an aggregation of 14 variables in these subindices, which better reflect trends that result from major political events and distinguish regional development patterns.

Alternative ways to measure inclusion were mentioned when adopting the participation-oriented point of view. In S-LCA, one can find other indicators such as Access to material resources, Access to immaterial resources and Community engagement. Affordability of goods and services to all people in the population was mentioned as a generic measure. This requires knowledge of the relative relation between what should be affordable and to what extent this needs financial sacrifice in the wallet of the individual. The Big Mac Index (as a specific affordability measure) was first quantified in 1987 by the Economist as the price difference of a burger from McDonalds in different countries to measure the purchasing power parity between countries' currencies [19]. Accessibility to resources was described as a means to measure the access to information, knowledge, and social groups among others. The participation rate describes the relative proportion of people engaged in an activity. Employment rate was mentioned and describes the extent that available labor resources are actually used (OECD 2023). The consideration rate was also mentioned and explained as a certain effectiveness of a policy to include stakeholders in its intended effects. In absence of a clear definition by an established institution, it was exemplified through criteria that assured the weight of certain stakes in the decision-making process. In the workshop, also the presence of formal recognition was mentioned. This had no established statistic but was meant to capture the degree to which certain activities or initiatives in a city are officially recognized. The level of institutionalization captures the extent of a formal status of an activity or initiative. The veil of ignorance is related to the ability of an individual to imagine the position of another [44]. The level of sharing was discussed as a means to consider how much of the time a product is shared and used [63].

Capturing inclusion with regard to UWMSs

Here too, indicators are categorized in terms of context and content.

Context indicators are meant to describe the socio-economic context of a city. Examples include the Inclusive Wealth Index and the Theil index. These indicators capture either a narrow economic scope (e.g., Palma ratio and Gini coefficient) or a broader societal scope (IHDI and MDI).

Content indicators describe measurements which can be influenced by local waste management policies. This category covers participatory aspects, for example, the presence of formal recognition of an organization or the consideration rate of marginalized groups during decision making. In this regard, inclusion indicators could be recognized for their differences in more active municipal influence (e.g. participation rate and level of sharing) or passive municipal influence (e.g. level of institutionalization and veil of ignorance).

Figure 4 shows the system boundaries, actors, resource flows, and indicators useful for the development of inclusive and circular UWMS. It shows the actors in the center as part of the formal waste sector with input and output flows (which can be described by the circular symbiosis indicators and circular actor indicators). The municipality is the dark grey layer that includes the formal sector. The municipality decides how it wishes to manage the formal sector by its inclusion-related policy indicators. The informal sector may be scoped outside of the municipalities control or consideration. Therefore, the informal sector is highlighted in light grey. The municipality manages the consideration of the informal sector by the inclusion-related context indicators. In essence, the circularity can be measured for this entire configuration of this UWMS system, through the proposed circularity context indicators, start and end of loop indicators.

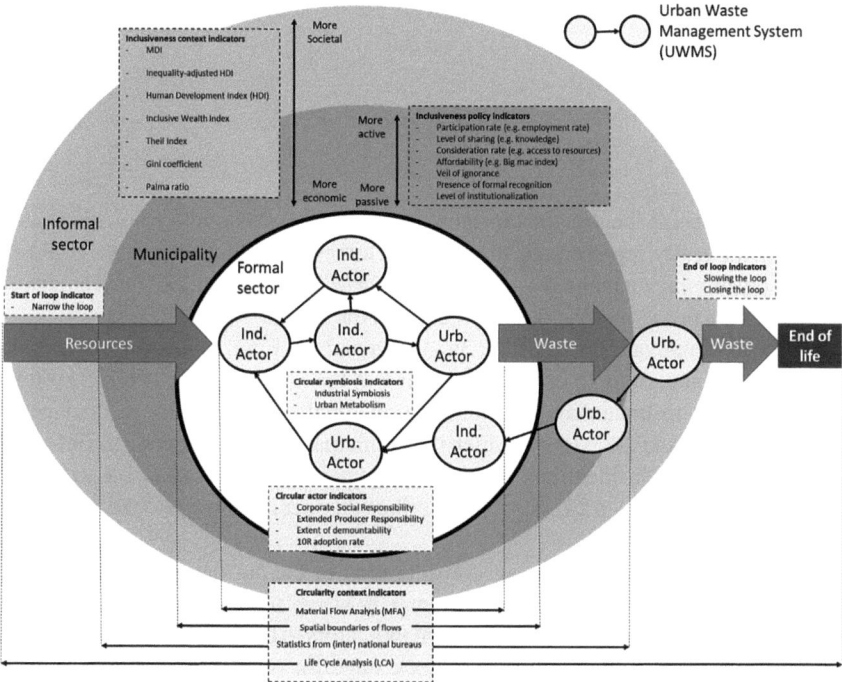

Fig. 4 System boundaries, actors, resource flows, and indicators useful for the development of inclusive and circular UWMSs

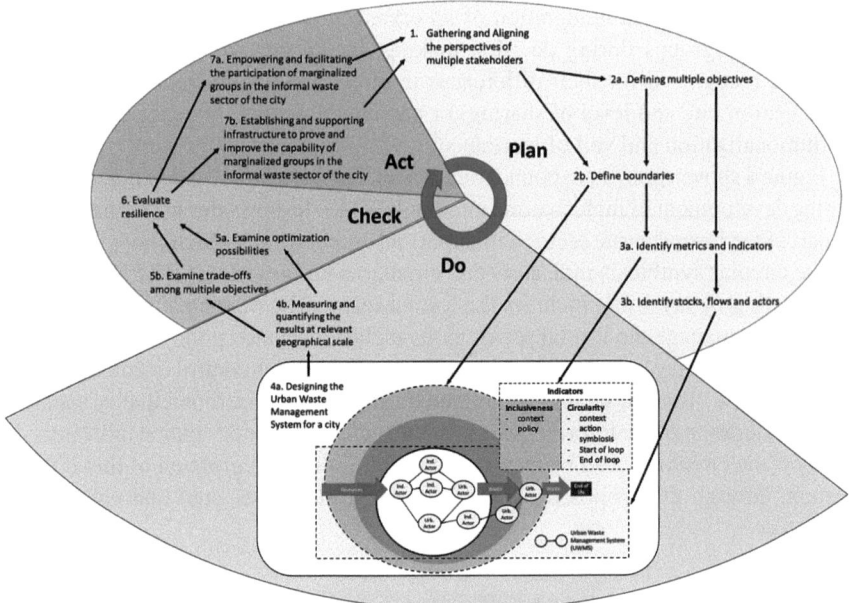

Fig. 5 Final design

3.4 Deliver

Here, we synthesize the previous insights from the workshops into a holistic framework schematically representing the various elements and indicators that could be used for developing an inclusive and circular UWMS. Figure 5 shows the framework integrating the process of the development and the design steps.

Synthesis is relevant for steps 2b, 3a, 3b, 4a and 4b. First, inputs for the UWMS need to be defined. Step 2b describes the system boundaries of the UWMS system. This is relevant to understand the jurisdiction for the municipality to manage its UWMS and to define the extent to which the municipality could integrate the informal waste sector. Step 3a identifies the relevant metrics and indicators based on the identified objectives (step 2a). Step 3b describes the specific flows, stocks, and actors that are part of the UWMS.

Step 4a allows the simultaneous consideration of the current actors, stocks, and flows, and how these make up the current UWMS system. Step 4b describes the relevant data to be collected and the relevant indicators to be measured and quantified. These results can then be fed into the check stage for optimization and trade-offs, and consequently for the evaluation of the regenerative potential of the UWMS.

4 Concluding Remarks

Both circularity and inclusion are important for characterizing any Urban Waste Management System (UWMS). Circularity indicators emphasize the monitoring of resources and waste flows running through a city and its UWMS. Inclusion may not only improve its circular capability, but it also boosts the social cohesion and regenerative potential of a city. Regeneration here is understood both in terms of increased availability of raw materials and in terms of agency provision to a plurality of stakeholders for collaborative learning and capacity building. The framework developed above offers a basis for improving UWMS with more complimentary solutions for circularity and inclusion allowing for further consideration of additional concepts that emerge in the literature such as that of the circular society [9, 10, 16].

The framework, however, has strong references to the Dutch waste management context which the authors of this chapter know best. Therefore, one critical limitation concerns the fact that different countries reveal different societal contexts, and these should be accounted for, *mutatis mutandis,* when applying it there. For example, the urbanization rate, the state of waste disposal, and type of environmental hazards can vary enormously across countries. Nonetheless the Double Diamond (DD) approach with its four sequential steps is, to our knowledge, largely universal in its applicability and many of the theoretical insights and indicator suggestions may need to be reconsidered, validated and calibrated in new institutional contexts and contents but will preserve their general applicability. The framework that emerges for this DD approach captures various aspects of circularity and inclusion in a city, yet it is agnostic towards the type of waste addressed. At the same time, it acknowledges that both dimensions have local as well as global consequences where the implications for different types of waste can be very different as they might be having their own sub-system configurations within an UWMS.

Appendix 1 Definitions of inclusiveness

Definition quotes	Authors	Year	Outlet
Inclusiveness is a **multi-dimensional concept** that needs to be unbundled before its connection with creativity is firmly established. Various tensions can arise when cities decide to adopt both creative city and inclusive city branding and urban policy initiatives	Alsayel, de Jong, Fransen	2022	Cities

(continued)

(continued)

Definition quotes	Authors	Year	Outlet
Principle of inclusiveness is the **idea** that healthcare staff and service users hold unique and valuable knowledge that can inform learning, as well as the notion that learning is a social process that involves people actively reflecting on shared knowledge. Despite initiatives to facilitate inclusiveness, research shows that embracing and learning from diverse perspectives is difficult	Kok, J., de Kam, D., Leistikow, I., Grit, K., Bal, R	2022	Health Care Analysis
Inclusiveness is multidimensional and comprised of spatial, social, environmental, economic, and political dimensions in which the characteristics of participation, equity, accessibility and sustainability are sometimes interwoven	Liang, D., De Jong, M., Schraven, D., Wang, L	2022	International Journal of Sustainable Development and World Ecology
The **principle** of inclusiveness in urban design which bears the meaning of "space for all" ensures the fulfillment of this aspect. Inclusiveness is one of the indicators of the Sustainable Development Goals (SDGs), and the same principles applies to Indonesia through the Decree of the Minister of Public Works number 5 year 2008, which includes the provision of green open spaces in order to launch Indonesian citizens' livelihood towards an ideal quality	Hariyani, D.S., Pratama, A.R	2021	IOP Conference Series: Earth and Environmental Science
Inclusiveness is a **key factor** for both job satisfaction and willingness to recommend, whilst innovation prone organizations appear to be the most attractive for nurses. The levers mix is slightly different among the age classes	Vainieri, M., Seghieri, C., Barchielli, C	2021	Health Services Management Research
Designing for inclusiveness is a complex and challenging activity that requires the adoption of a specific inclusive design process to create human-centred solutions based on user desire and needs	Canina, M.R., Parise, C., Bruno, C	2021	Lecture Notes in Networks and Systems

(continued)

(continued)

Definition quotes	Authors	Year	Outlet
Inclusiveness is largely shaped by domestic politics, pre-existing institutions and power relations, as well as the resources, capacities and prior experiences of civil society and subnational governments. In practice, the SDGs' emphasis on inclusiveness does not necessarily mean that a wider range of perspectives are taken into account in domestic contexts	Siegel, K.M., Bastos Lima, M.G	2020	World Development
Inclusiveness is a **key theme** in scholarship on Responsible Innovation (RI) and Responsible Research and Innovation (RRI). RI/RRI researchers make a strong case for involving stakeholders into science and innovation processes	Koch, S	2020	Journal of Responsible Innovation
Inclusiveness is conducive to expanding the connotation of urbanization quality, which could provide guidance for improving the quality of urbanization	Yu, W., Zhao, L	2018	Chinese Journal of Applied Ecology
Inclusiveness is essential to deliberative democracy, but factors influencing citizens' willingness to participate in deliberation need to be better understood	Jennstål, J	2018	European Political Science Review
Inclusiveness is the **process** of valuing, respecting and supporting members of an entity. Resilience in permanent organizations can be defined as the capability to respond to and prepare for disruption and thus, promote business continuity	Blay, K	2018	Construction Research Congress
The **notion** of dynamic inclusiveness is framed in terms of imagined normative allocations of the inter-temporal product of growth, as dictated by notions of equity of varying orders of demandingness	Jayaraj, D., Subramanian, S	2012	Economic and Political Weekly
Inclusiveness is the **growth mantra** for policymakers today - not just in India, but the world over	Das, K	2010	Journal of Rural Development
Inclusiveness is a **way** to increase efficiency in city management and service delivery across urban and peri-urban areas. New norms of practice and reform of institutional procedures in cities is dependent upon strengthening modes of inclusive urban governance	McCarney, P	2010	Peri-urban Water and Sanitation Services: Policy, Planning and Method

(continued)

(continued)

Definition quotes	Authors	Year	Outlet
Inclusiveness is an important **principle** of a country's democracy and can encourage citizens to participate in politics	Kittilson, M.C., Schwindt-Bayer, L	2010	Journal of Politics
By inclusiveness is meant the **existence of mechanisms**, formal or informal, to extend terms and conditions negotiated by workers with strong bargaining power to workers with less bargaining power	Salverda, W., Mayhew, K	2009	Oxford Review of Economic Policy
Inclusiveness is a matter of definition and **process** that has been encountered in other foresight style activities where the opinions of the polity need to be taken into account	Loveridge, D., Street, P	2005	Foresight
Based on a historical review of successive conceptualizations of development, his case for "inclusiveness" is a **plea** for correcting this asymmetrical process striking a better balance between economic efficiency, decent work and environmental protection	Sachs, I.2004	2004	International Labour Review
At the heart of optimal distinctiveness theory is the idea that a group's level of inclusiveness is a **significant determinant** of how well that group can meet members' needs for assimilation and differentiation	Pickett, C.L., Silver, M.D., Brewer, M.B	2002	Personality and Social Psychology Bulletin
Using SASA as the basis for our discussion, we argue that inclusiveness is inextricably tied to discourses about democracy, which privilege the notion of participation, through which it is assumed inclusiveness will be achieved in ways that are considered to be appropriate. We demonstrate that rather than actually realizing the full extent of inclusiveness made possible by the new Constitution, SASA circumscribes such inclusiveness in ways that may potentially marginalize the historically marginalized in South Africa (inter alia, black, working-class and rural people), and rather than redressing past inequities, may perpetuate and further exacerbate them	Sayed, Y., Carrim, N.1998	1998	International Journal of Inclusive Education

(continued)

(continued)

Definition quotes	Authors	Year	Outlet
Inclusiveness is increased when teachers expose students to: (a) historical theories that demonstrate early respect for diversity and the role of sociocultural events in personality formation, (b) recent reformulations of traditional theory, (c) feminist approaches that focus on the relational elements of development, and (d) cognitive developmental perspectives that value desires for related-ness and strivings for independence	Enns, C.Z	1989	Teaching of Psychology

References

1. Anttiroiko A, De Jong M (2020) The inclusive city; the theory and practice of creating shared urban prosperity. *Springer eBooks*
2. Aparcana S, Salhofer S (2013) Development of a social impact assessment methodology for recycling systems in low-income countries. Int J Life Cycle Assess 18(5):1106–1115
3. Ardi R, Leisten R (2016) Assessing the role of informal sector in WEEE management systems: a system dynamics approach. Waste Manage 57:3–16
4. Bastos J, Garcia R, Freire F (2019) Indicators for waste prevention and management—measuring circularity. https://www.urbanwins.eu/wp-content/uploads/2019/05/d2.3-Indicators-for-waste-prevention-and-management.pdf
5. Bocken N (2021) Sustainable business models. In Leal Filho T, Azeiteiro W, Azul U, Brandli AM, Özuyar L, Wall P (ed) Decent work and economic growth. Encyclopedia of the UN sustainable development goals. Springer
6. Bocken N, Miller K, Evans S (2016) Assessing the environmental impact of new Circular business models. In Jonker J, Faber NR (eds.) The proceedings of the first international conference on "New Business Models": exploring a changing view on organizing value creation (pp. 17–18). Toulouse Business School
7. Bocken N, Strupeit L, Whalen K, Nußholz J (2019) A review and evaluation of circular business model innovation tools. Sustainability (Switzerland) 11(8):1–25
8. Brändström J, Saidani M (2022) Comparison between circularity metrics and LCA: a case study on circular economy strategies. J Clean Prod 371(July)
9. Calisto Friant M, Vermeulen WJV, Salomone R (2020) A typology of circular economy discourses: navigating the diverse visions of a contested paradigm. Resour Conserv Recycl 161(April):104917
10. Calisto Friant M, Vermeulen WJV, Salomone R (2021) Analysing European Union circular economy policies: words versus actions. Sust Prod Cons 27:337–353
11. Campitelli A, Kannengießer J, Schebek L (2022) Approach to assess the performance of waste management systems towards a circular economy: waste management system development stage concept (WMS-DSC). MethodsX 9(February):101634
12. Campitelli A, Schebek L (2020) How is the performance of waste management systems assessed globally? A systematic review. J Clean Prod 272(July)
13. Census (2021) Theil Index. Accessed on 28th of March 2023 at https://www.census.gov/topics/income-poverty/income-inequality/about/metrics/theil-index.html#:~:text=The%20Theil%20index%20is%20a,everyone%20having%20the%20same%20income

14. Céspedes Restrepo JD, Morales-Pinzón T (2018) Urban metabolism and sustainability: precedents, genesis and research perspectives. Resour Conserv Recycl 131(16):216–224
15. Cramer J (2020) How network governance powers the circular economy. In Amsterdam Economic Board. www.amsterdameconomicboard.com
16. Dewick P, de Mello AM, Sarkis J, Donkor FK (2022) The puzzle of the informal economy and the circular economy. Resour Conserv Recycl 187(August):106602
17. Dias SM (2016) Waste pickers and cities. Environ Urban 28(2):375–390
18. Dörffel C, Schuhmann S (2022) What is inclusive development? Introducing the Multidimensional Inclusiveness Index. In *Social Indicators Research* (Issue 0123456789). Springer Netherlands
19. Economist (1998) Big MacCurrencies. https://www.economist.com/finance-and-economics/1998/04/09/big-maccurrencies
20. Elia V, Gnoni MG, Tornese F (2017) Measuring circular economy strategies through index methods: a critical analysis. J Clean Prod 142:2741–2751
21. Ellen MacArthur Foundation (2019) The circular economy in detail. Www.Ellenmacarthurfoundation.Org. https://archive.ellenmacarthurfoundation.org/explore/the-circular-economy-in-detail
22. European Commission (2021) Circular material use rate. Ec.Europa.Eu. https://ec.europa.eu/eurostat/web/products-datasets/-/cei_srm030
23. Friant MC, Vermeulen WJV, Salomone R (2020) Exploring four visions of a circular future: from Technocentric Circular Economy to Transformational Circular Society (Issue November). https://www.researchgate.net/project/CRESTING-Circular-Economy-Sustainability-Implications-and-Guiding-Progress
24. Geissdoerfer M, Pieroni MPP, Pigosso DCA, Soufani K (2020) Circular business models: a review. J Clean Prod 277:123741
25. Girardet H (2015) Creating regenerative cities. Routledge
26. Goodwin Brown E, Sosa L, Schröder A, Bachus K, Bozkurt Ö (2020) The social economy—a means for inclusive and decent work in the circular economy? https://www.circle-economy.com/resources/the-social-economy-a-means-for-inclusive-decent-work-in-the-circular-economy
27. Gutberlet J (2021) Grassroots waste picker organizations addressing the UN sustainable development goals. World Dev 138:105195
28. Gutberlet J, Carenzo S (2020) Waste pickers at the heart of the circular economy: a perspective of inclusive recycling from the Global South. Worldwide Waste J Interdisc Stud 3(1):6
29. Haberl H, Wiedenhofer D, Pauliuk S, Krausmann F, Müller DB, Fischer-Kowalski M (2019) Contributions of sociometabolic research to sustainability science. Nat Sust 2(3)
30. Kirchherr J, Reike D, Hekkert M (2017) Conceptualizing the circular economy: an analysis of 114 definitions. Resour Conserv Recycl 127:221–232
31. Konietzko J, Das A, Bocken N (2023) Towards regenerative business models: a necessary shift? Sustainable Production and Consumption 38:372–388
32. Liang H, Renneboog L (2016) On the foundations of corporate social responsibility. J Financ 72(2):853–910
33. Liu Z, Schraven D, De Jong M, Hertogh M (2023) The societal strength of transition: a critical review of the circular economy through the lens of inclusion. Int J Sust Dev World 30(7):826–849
34. Lombari DR, Laybourn P (2012) Redefining industrial symbiosis. J Ind Ecol 16(1):28–27
35. López de Munain D, Castelo B, Ruggerio CA (2021) Social metabolism and material flow analysis applied to waste management: a study case of autonomous city of Buenos Aires, Argentina. Waste Manage 126:843–852
36. Makriyannis C (2022) The foundational economy-as-an-organism assumption of ecological economics: is it scientifically useful? Ecol Econ 200:107541
37. Maalouf A, Mavropoulos A, El-Fadel M (2020) Global municipal solid waste infrastructure: delivery and forecast of uncontrolled disposal. Waste Manage Res 38(9):1028–1036

38. Mulder K (2016) Urban symbiosis: a new paradigm in the shift towards post-carbon cities. NewDist, (July), 16–24
39. OECD (2023) Income inequality. Accessed on 28th of March 2023 at https://data.oecd.org/inequality/income-inequality.htm
40. OECD (2023) Employment rate. Accessed on 28th March of 2023 at https://data.oecd.org/emp/employment-rate.htm
41. Paiho S, Mäki E, Wessberg N, Paavola M, Tuominen P, Antikainen M, Heikkilä J, Antuña C, Jung N (2020) Towards circular cities—conceptualizing core aspects. Sustain Cities Soc 59(March):102143
42. Paiho S, Wessberg N, Pippuri-Mäkeläinen J, Mäki E, Sokka L, Parviainen T, Nikinmaa M, Siikavirta H, Paavola M, Antikainen M, Heikkilä J, Hajduk P, Laurikko J (2021) Creating a circular city–an analysis of potential transportation, energy and food solutions in a case district. Sust Cities Soc 64(September 2020), 102529
43. Potting J, Hanemaaijer A, Delhaye R, Ganzevles J, Hoekstra R, Lijzen J (2018) Circular economy: what we want to know and can measure. Framework and baseline assessment for monitoring the progress of the circular economy in the Netherlands. *PBL Policy Report*. *PBL Publicaiton Number*, *3217*, 92. https://circulareconomy.europa.eu/platform/sites/default/files/pbl-2019-outline-of-the-circular-economy-3633.pdf
44. Rawls J (2001) Justice as fairness: a restatement. Belknap Press, Cambridge, Massachusetts, p 2001
45. Scheinberg A, Anschütz J, Van de Klundert A (2006) *Waste pickers: poor victims or waste management professionals?*
46. Scheinberg A, Nesic J, Savain R, Luppi P, Sinnott P, Petean F, Pop F (2016) From collision to collaboration - integrating informal recyclers and re-use operators in Europe: a review. Waste Manage Res 34(9):820–839
47. Schröder P (2020) Promoting a just transition to an inclusive circular economy. *The Royal Institute of International Affairs, Chatham House*, April, 1–33
48. Shreeves AD (2020) The significance of the informal waste sector in a minority world country: A case-study of metropolitan Atlanta [MSc thesis, Georgia State University]. https://scholarworks.gsu.edu/cgi/viewcontent.cgi?article=1155&context=geosciences_theses
49. Singh SJ, Talwar S, Shenoy M (2021) Why Socio-metabolic studies are central to ecological economics. Ecol Econ Soc INSEE J 4(2):21–43
50. Steuer B (2021) Hunting for hidden treasures: a research methodology on China's informal recycling sector. The Routledge Handbook of Waste Studies, January, 154–168
51. Subagio H, Santosa RE, Setiawan MI (2020) Community behavior, regulation, and reliable waste infrastructure in ngawi regency to improve the quality of life. In *Proceedings of the International Conference on Industrial Engineering and Operations Management* (Vol. 59, pp. 2920–2930)
52. Temper L, Bene D del, Martinez-Alier J (2015) *Mapping the frontiers and front lines of global environmental justice: The EJAtlas*. J Pol Ecol
53. Thomson G, Newman P (2018) Urban fabrics and urban metabolism—from sustainable to regenerative cities. Resour Conserv Recycl 132:218–229
54. Tong X, Tao D (2016) The rise and fall of a "waste city" in the construction of an "urban circular economic system": The changing landscape of waste in Beijing. Resour Conserv Recycl 107:10–17
55. Tong X, Yu H, Han L, Liu T, Dong L, Zisopoulos FK, Steuer B, de Jong M (2023) Exploring business models for social inclusiveness and carbon emission reduction via post-consumer recycling infrastructures in China: An agent-based modelling approach. Res Conserv Recycl 188(August 2022), 106666
56. Tong X, Yu H, Liu T (2021) Using weighted entropy to measure the recyclability of municipal solid waste in China: exploring the geographical disparity for circular economy. J Clean Prod 312:127719
57. Tong X, Yu H, Liu T (2021) Using weighted entropy to measure the recyclability of municipal solid waste in China: exploring the geographical disparity for circular economy. J Clean Prod 312(April):127719

58. United Nations Environment Programme (UNEP) (2022) *Why take action*. Www.Buildingcircularity.Org. https://buildingcircularity.org/
59. Williams J (2021) Circular cities: what are the benefits of circular development? Sustainability 13(10):5725
60. Wilson DC, Velis CA, Rodic L (2013) Integrated sustainable waste management in developing countries. Proc Inst Civil Eng Waste Res Manag 166(2):52–68
61. World Bank (2023) Metadata glossary. Accessed on 28th of March at https://databank.worldbank.org/metadataglossary/gender-statistics/series/SI.POV.GINI
62. World Econonic Forum & PwC (2018) *Circular economy in cities—evolving the model for a sustainable urban future*. https://is4ie.org/resources/documents/28
63. Xu X (2020) How do consumers in the sharing economy value sharing? Evidence from online reviews. Dec Supp Syst 128, January 2020, 113162
64. Zisopoulos FK, Schraven DFJ, de Jong M (2021) How robust is the circular economy in Europe? An ascendency analysis with Eurostat data between 2010 and 2018. Res Conserv Recycl 178(106032)
65. Zisopoulos FK, Noll D, Singh SJ, Schraven D, De Jong M, Fath BD, Goerner S, Webster K, Fiscus D, Ulanowicz RE (2023) Regenerative economics at the service of islands: assessing the socio-economic metabolism of Samothraki in Greece. J Clean Prod 408:137136

Open Access This chapter is licensed under the terms of the Creative Commons Attribution 4.0 International License (http://creativecommons.org/licenses/by/4.0/), which permits use, sharing, adaptation, distribution and reproduction in any medium or format, as long as you give appropriate credit to the original author(s) and the source, provide a link to the Creative Commons license and indicate if changes were made.

The images or other third party material in this chapter are included in the chapter's Creative Commons license, unless indicated otherwise in a credit line to the material. If material is not included in the chapter's Creative Commons license and your intended use is not permitted by statutory regulation or exceeds the permitted use, you will need to obtain permission directly from the copyright holder.

Managing the Transition to a Circular Urban Waste Management System

Afsaneh Moradi

Abstract Transition to a circular urban waste management system is a complex, incremental process involving systemic change. Transition studies have focused on identifying actions and plans to support this radical shift, with governance and control of the process primarily following the transition management approach. Transition management encompasses strategic, tactical, operational, and reflexive activities that are designed to guide, manage, and lead the process of systemic change. In the Netherlands, managing the transition towards a circular urban waste management system has become a priority for cities seeking to address environmental and economic challenges associated with traditional linear waste management practices. This chapter examines the principles of transition management in the context of transforming urban waste management system into a circular and more resilient system. It explores the different types of governance activities involved in this process and investigates how existing capabilities can be leveraged to support the transition toward circular urban waste management practices. Ultimately, by reviewing the current practices of circular urban waste management in the Netherlands, this chapter aims to provide insights to inform policy and practice in managing the transition process, both at the national and international levels.

Keywords Transition management · Systemic change · Governance · Urban waste · Circularity · Projects · Policy accumulation

1 Introduction

The problem of waste disposal emerged as a result of rapid urbanization and increasing population, especially in cities. In the late 1980s, the traditional waste management system that was based on landfilling, composting, and incineration were no longer adequate to meet the needs of ever-growing cities and urban areas,

A. Moradi (✉)
Department of Organization, Strategy, and Entrepreneurship, School of Business and Economics, Maastricht University, Tongersestraat, 53, 6211 LM, Maastricht, The Netherlands
e-mail: a.moradi@hmaastrichtuniversity.nl

© The Author(s) 2026
M. de Jong et al. (eds.), *The Inclusive Circular Economy*, Urban Sustainability,
https://doi.org/10.1007/978-981-96-6867-0_5

mainly due to the lack of landfill capacity and insufficient thermal treatment capacity. This highlighted the need of holistic changes in the waste management system; a fundamental shift is required to disrupt the whole system's functions, initiate, and guide long-term transformational change [19]. These fundamental shifts are often referred to as systemic changes, that are performed through strategic planning, design thinking, promoting new business models, and responsible consumption and production. It often involves a significant rethinking of visions and goals and requires changes in the structures and business practices. In this process, the system needs to be evolved with the collaboration of many teams and change agents.

The transition management model is a systemic approach that aims to initiate and guide long-term transformational change in complex systems. This model emphasizes collaboration and engagement among various stakeholders, including government, waste management companies, consumers, and other relevant organizations. According to [22], transition management is a governance approach based on insights from governance and complex systems theory and practical experimentation and experience. The transition management framework distinguishes between different types of governance activities that influence long-term change, and is used both to "analyze and structure", or "manage" ongoing governance processes in society. The framework, identifies four types of governance activities relevant to societal transitions: strategic, tactical, operational, and reflexive.

In 1975, the Dutch government established the Waste Disposal Act (Wet op de Afvalstoffen), which laid down the legal framework for waste management in the country. The Act introduced measures to reduce waste, promote recycling and reuse, and regulate the disposal of hazardous waste, but that wasn't enough, and as stated by the Rijkswaterstat[1] in the late 80s, the Netherlands suffered from a lack of landfill capacity and insufficient thermal treatment capacity. Therefore, there was a need for more ambitious goals and gradual changes.

The most significant changes to the Dutch waste management system began in the 1990s, with the introduction of the National Waste Management Plan (Landelijk Afvalbeheer Plan, LAP1) in 1992. LAP1 was designed to provide a comprehensive framework for managing waste in the country, with a particular focus on reducing waste and increasing recycling. The plan included targets for reducing the amount of waste going to landfill and increasing the recovery of materials from waste streams. LAP1 was followed by LAP2 in 1997 and LAP3 in 2009, which continued to build on the principles and goals set out in the original plan. The results of these changes, as stated by Rijkswaterstat, were a decrease in the amount of landfilled waste from 35% in 1985 to 2.1% in 2016, raising the rate of recovery (including Waste-to-energy) from 50 to 93%, and improved separation approaches [40].

Today, 77% of waste is recycled, and the residual waste is mainly used for energy production [40]. The Netherlands continues to update and refine its waste management strategies to achieve its ambitious sustainability goals, by emphasizing the importance of the transition to a circular economy that offers economic opportunities by being less dependent on the import of scarce raw materials. This chapter aims

---

[1] The executive organization of Ministry of infrastructure and water management.

to show how different steps of the transition management model are applied to the waste management system in the Netherlands.

This chapter is structured as follows: Sect. 2 presents the concept of transition to a circular waste management system in the Netherlands, which primarily explains the plans and programs developed in this context. Section 3 introduces the concept of transition management process and its relevant steps, and provides evidence of the process management in the Netherlands based on the defined management model. Finally, Sect. 4 presents the concluding remarks.

2 Transition Toward Circular Waste Management System in the Netherlands

The transition towards a circular waste management system can help reduce the amount of waste that goes to landfills and incinerators, and can promote a more sustainable and environmentally friendly way of managing waste.

The Netherlands has been at the forefront of implementing circular waste management practices by setting ambitious targets for reducing, reusing, and recycling different types of waste. The country has a well-established waste management system that emphasizes the importance of reducing waste, promoting recycling and circularity, and ensuring responsible waste disposal. The main goal of the circular economy is to use the lowest possible amount of virgin raw material, using and reusing resources sustainably and efficiently, in order to minimize waste. This vision is articulated in the Dutch government's National Waste Management Plan (Landelijk Afvalbeheer Plan), which is the policy framework for waste in the circular economy in the Netherlands. The third LAP, (LAP3), supports the transition to a circular economy [36].

LAP3 sets out ambitious targets for reducing waste, and increasing the recovery of materials from waste streams. Specifically, the plan aims to achieve a recycling rate of 75% for municipal solid waste by 2020, and to reduce the amount of residual waste going to landfills to 100 kg per capita per year by 2020. In addition, LAP3 promotes the development of new technologies and business models that support a circular economy, such as product design for recyclability, and the use of waste as a resource for energy generation and material production. According to the latest available data, the Netherlands did not achieve its target of a 75% recycling rate for municipal solid waste by 2020 (Rijksoverheid, n.d.). The recycling rate for municipal waste in the Netherlands was 56.9% in 2020 [11], which is the third highest rate in Europe but still below the 75% target set in the National Waste Management Plan (LAP3).

The other program for a transition to a sustainable and circular economy is the government-wide program 'The Netherlands Circular in 2050', launched in September 2016. It is the overarching vision for the circular economy in the Netherlands and sets the goal for a fully circular economy by 2050. This vision is supported by the National Agreement on the Circular Economy (NACE) and the LAP series, as they both contribute to the implementation of circular economy principles at the local and national levels. The first milestone of this plan was the "raw material agreement", signed in January 2017 by 180 parties from both government and industry actors who agreed to reduce the use of primary raw materials (minerals, fossils, and metals) by half until 2030 [16]. To achieve this target the government set out three specific goals:

1. Raw materials in existing supply chains are utilized in an efficient and high-quality manner.
2. In cases in which new raw materials are needed, fossil-based, critical, and non-sustainably produced raw materials are replaced by sustainably produced, renewable, and generally available raw materials when possible.
3. New production methods and products will be designed for a circular economy, areas will be reorganized, and new ways of consumption will be promoted in order to give an extra boost to the desired reduction, replacement, and utilization of raw materials for strengthening the economy [27].

The program highlighted the collaboration of different parties in the public and private sectors including the government, companies, social organizations, NGOs, and knowledge centers. The National Agreement on the Circular Economy (NACE) is a cooperative agreement signed in 2018 between the Dutch government, businesses, NGOs, and research institutions aimed at accelerating the transition to a circular economy in the Netherlands. This agreement outlines a series of actions and commitments to achieve a fully circular economy by 2050.

Partners involved in the National Agreement on the Circular Economy (NACE) in the Netherlands develop specific plans and strategies, called "transition agendas," in 2018, to achieve the targets set out in the agreement. These transition agendas are informed by advisory reports from the Social and Economic Council (SER) and the Council for the Environment and Infrastructure (Rli).[2] In the context of the circular economy, these bodies have provided reports on topics such as circular procurement, sustainable agriculture, and the role of cities in the transition to a circular economy. The partners involved in the National Agreement on the Circular Economy use these reports to inform their transition agendas, which will outline specific actions and measures to achieve the targets set out in the agreement. Figure 1 shows the timeline for the transition to a circular economy and the related programs in the period of 2016 to 2050.

---

[2] The SER and Rli are advisory bodies that provide recommendations to the Dutch government on social, economic, and environmental issues.

Fig. 1 Timeline for the transition to a circular economy during the period of 2016 to 2050. Developed based on the data from [16]

3 Managing Transition Process

Managing a transition process can be a challenging task, but with the right approach, it can be executed smoothly. The Transition Management Cycle (TMC) is a framework for managing sustainability transitions, developed by [23]. The TMC is a cyclical process that involves various stages and steps to facilitate sustainability transitions. This model provides a comprehensive approach to sustainability transitions, encompassing strategic, tactical, operational, and reflexive phases. In this section, we will explore how the TMC has been applied to facilitate the circular waste management transition in the Netherlands, focusing on the steps of the TMC and how they were implemented in practice.

- **Strategic activities**: The strategic phase of the TMC consists of problem-structuring and envisioning activities, which set the direction and vision for the transition. Problem structuring involves analyzing the issue at hand and breaking it down into its parts to better understand it. This process also involves analyzing the underlying causes of problems, including societal values, economic incentives, and technological limitations [45]. Another strategic activity is envisioning, which involves creating a shared vision of a more sustainable future [23]. The envisioning process engages stakeholders in a collaborative process to develop a shared vision of the future, and establish a direction for the transition [41].
- **Tactical activities:** The tactical phase of TMC includes agenda-setting and coalition-forming activities, which are critical for building the necessary political and social support for the transition. Agenda-setting involves identifying the key issues that need to be addressed and setting priorities [51]. The TMC is a participatory and process-oriented approach, enabling stakeholders to work together towards a sustainable future by engaging in ongoing dialogue and feedback [3]. Coalition-forming involves bringing together stakeholders who share a common interest in the issue and building a network of support [12]. The iterative process of the TMC allows for feedback and adjustment, which is crucial for building consensus and addressing competing interests among stakeholders [25].
- **Operational activities:** The operational phase of the TMC involves experimenting with new solutions and creating support structures for the transition. This phase is critical for putting the vision and agenda into action. Operational activities include designing and testing new solutions, such as innovative waste management systems, and creating supportive structures, such as circular economy knowledge platforms. The TMC emphasizes the importance of learning through experimentation and feedback, which enables stakeholders to refine and improve their approaches to sustainability [18]. This iterative process helps to build capacity and overcome barriers to implementation, such as regulatory and financial constraints [24].
- **Reflexive activities:** The reflexive phase of the TMC involves closely monitoring the transition process, and reflecting on the outcomes to inform future actions. This phase is critical for evaluating the effectiveness of the solutions implemented and identifying areas for improvement. Reflexive activities include evaluating the social, economic, and environmental impacts of the transition process, and identifying opportunities for further innovations [1]. The TMC emphasizes the importance of ongoing learning and feedback, which enables stakeholders to continuously improve their approaches to sustainability [22]. This iterative process helps to build resilience and adaptive capacity, both of which essential for addressing complex sustainability challenges [49].

Based on the definitions of the transition management process provided, the subsequent sections of this chapter will examine the practical elements of the circular waste management system in the Netherlands. These sections will illustrate how the Dutch government has transformed the waste management system into a sustainable and circular one that aims to reduce waste, reuse materials, and recycle resources while minimizing the negative impact on the environment and promoting a more sustainable and circular economy.

3.1 Strategic Activities: Setting the Transition Arena

Loorbach [23] defined the strategic activities involved in societal (sub-) systems as vision development, strategic discussions, long-term goal formulation, collective goal and norm-setting, and long-term anticipation [23]. The literature on transition management and sustainable development governance emphasizes the importance of strategic long-term planning and inclusive governance, which involves developing a shared vision, creating a governance network, experimenting with new approaches, and involving multiple stakeholders in policymaking [2, 32]. While this may require a shift in traditional policymaking approaches, it is necessary for creating a more sustainable future.

The key problems with the waste management system in the Netherlands, such as high levels of waste generation, limited recycling and reuse, and environmental degradation, have been identified over several decades. The Dutch government has been working to address these issues since the 1980s, when they first introduced policies to promote waste separation and recycling. Over time, the government has implemented a range of measures to incentivize waste reduction and recycling, such as landfill taxes, extended producer responsibility, and the promotion of circular business models. Despite these efforts, however, the Netherlands still faces challenges related to waste management, such as a high per capita waste generation rate (515 kg per capita) [10], and relatively low recycling rates for certain waste streams. As a result, the Dutch government has continued to identify and address these problems through ongoing policy development and stakeholder engagement.

After identifying the main problems of the waste management system, the Dutch government has set an ambitious goal of achieving a fully circular economy by 2050 through its National Raw Materials Agreement, which is a collaborative effort between the government, industry, and civil society. This agreement introduces a key area of focus, which is the development of a circular waste management system to reduce waste and increase the use of secondary raw materials.

The Dutch waste management system is influenced by the waste hierarchy (Lansnik's ladder), which was introduced into Dutch legislation in 1994 and later incorporated into the European Waste Framework Directive. The ladder's basic principles include reducing waste generation, recovering valuable raw materials from waste, generating energy by incinerating residual waste, and landfilling only what's left in an environmentally sound way [20]. The Netherlands introduced a landfill tax in 1995 to reduce waste generation by making waste disposal more expensive, and at the same time promoting recycling and incineration as more attractive waste management options. Instruments like the landfill tax and volume-based waste fees facilitate the shift towards less landfilling and increased waste recovery and recycling.

To create a shared vision for a circular waste management system, the Dutch government launched a government-wide program for a Circular Dutch Economy by 2050 in 2016, and outlines the necessary steps towards achieving a sustainable, fully circular economy by the year 2050 [38]. The vision sets the long-term goals and objectives for the system and provides a common understanding among stakeholders. For example, it provides a roadmap for using raw materials, products, and services in a smarter and more efficient way to reduce waste and promote environmental sustainability. This vision has been supported by various policy documents, including the "National Waste Management Plan" and the "Dutch Circular Economy Agenda", which have provided a framework for waste management and circular economy initiatives [5].

Developing a vision for a circular waste management system requires a comprehensive and coordinated approach that involves a wide range of stakeholders, including government agencies, businesses, and individuals. To bring the stakeholders together and facilitate their collaborations, the "Green Deal programme" was launched in October 2011 by the Netherlands Ministry of Economic Affairs with the joint aims of saving energy, materials, and water. This initiative has refined the role of government from directive policymaker to responsive service provider supporting organizations that seek assistance in realizing the circular economy opportunities aligned with these aims, and facing implementation barriers [9]. The government also conducted consultations with stakeholders from various sectors, including businesses, NGOs, and citizens. These consultations aimed to identify challenges and opportunities in waste management process, and develop a shared understanding of the goals and priorities of a circular economy [34].

3.2 *Tactical Activities: The Transition Agenda*

The tactical activities, as defined by [23], include interest-driven steering activities that are related to the dominant structures (regimes). This consists of all established patterns and structures, such as rules and regulations, institutions, organizations and networks, infrastructure, and routines.

The Netherlands, like many other countries, faces a significant challenges in managing waste sustainably. As the world's population continues to grow, and consumption patterns change, the amount of waste generated is increasing at an alarming rate. This presents a significant environmental and economic challenge, as traditional linear waste management systems are unsustainable in the long run.

After strategic planning and setting the transition arena, the focus shifted to defining the transition agenda. It is critical to take practical steps toward achieving the defined goals for implementing sustainable circular waste management practices. Key priorities include reducing waste at the source, promoting recycling and upcycling technologies, and establishing circular supply chains.

To define the transition agenda, after signing the "Raw material agreement" in 2017, the Dutch government and its signatories developed five transition agendas in 2018, with a focus on five sectors and value chains that are economically significant but also have a high environmental burden. The five transition agendas are:

- Plastics,
- Consumer Goods,
- Manufacturing,
- Construction,
- Biomass and Food

The agendas set out how the relevant sector can become circular by 2050, and what actions need to be taken [35]. Besides this, the national waste management plan describes the main practical steps for waste prevention, separation, collection, transportation, trade and mediation, reusing, recycling, and the incineration or landfilling the residual waste. This approach is commonly referred to as the 'order of preference' which defines a hierarchy of waste management options for prioritizing waste reduction, reuse, and recycling over landfilling or incineration. The order of preference is as follows:

3.2.1 Waste Prevention

The first priority in the waste management hierarchy is to prevent waste from being generated. This can be achieved through measures such as designing products for durability, reducing packaging, and promoting sustainable consumption patterns. One of the main actions in this context is Extended Producer Responsibility (EPR).

The OECD defined EPR as "the producer's responsibility for a product to be extended to the post-consumer stage of a product's life cycle" [30]. The Netherlands has implemented an EPR system, which makes producers responsible for the entire life cycle of their products, including the management of their end-of-life phase. This system incentivizes producers to design products that are more easily recyclable, reusable, or repairable. The EPR system contributed significantly to promoting sustainable waste management practices and reducing waste in the Netherlands. It requires producers or importers who introduce products into the market from the five categories[3] mentioned above, to take responsibility for the entire life cycle of those products. This entails the establishment of a waste management system for products, as well as the arrangement of financing for their disposal. Such responsibility is part of a chain-responsibility approach, in which other stakeholders, such as municipalities and retailers, are also responsible for managing the waste generated in their respective operations. According to the Dutch government's Point of Single Contact for entrepreneurs (business.gov.nl), there are plans to extend the requirements for

---

[3] In the Netherlands, producer responsibility (in Dutch: producentenverantwoordelijkheid) is in place for 5 products: electrical and electronic equipment, batteries and accumulators, scrap vehicles, car tires, packaging.

existing products to which producer responsibility applies, as well as to broaden the range of products.[4] There are two key elements of EPR design in the Netherlands:

- The mix of EPR instruments used, and
- Whether producers fulfill their obligations individually or collectively.

The most common EPR instruments are take-back requirements, advance disposal and recycling fees, and deposit-refund systems [8].

3.2.2 Waste Separation

Waste separation is a crucial tactical activity within the circular waste management system in the Netherlands. The Dutch government has implemented various policies and initiatives to promote the separation of waste at the source, encouraging households and businesses to sort their waste into different categories, such as paper, plastic, organic waste, and residual waste. The separated waste is then collected and transported to specialized treatment facilities, where it is processed for recycling or energy recovery. This approach enables the recovery of valuable materials and reduces the amount of waste sent to landfills.

Research has shown that waste separation can significantly reduce the amount of waste generated and increase the recovery of valuable resources. For instance, a study by the World Bank found that improved waste management practices, including waste separation, could reduce global solid waste generation by up to 20%, with a significant impact on greenhouse gas emissions [4]. Currently, separating waste at the source is a more efficient method compared to mechanical post-consumer separation. When waste fractions are disposed of together, they tend to contaminate each other, making material recovery more difficult and less effective. However, the effectiveness of source separation relies heavily on the participation and waste separation practices of residents.

The Dutch government has implemented several initiatives to promote waste separation, including education and awareness campaigns, the provision of recycling facilities in public areas, and financial incentives for households and businesses to participate in waste separation programs. There is an established successful source separation system for household waste, which has proven effective in increasing recycling rates and supporting the country's shift toward a more circular waste management model.

Overall, it can be argued that waste separation is an essential component of the circular waste management system in the Netherlands. It enables the recovery of valuable resources, reduces landfill dependency, and promotes a more sustainable and circular approach to waste management.

---

[4] https://business.gov.nl/running-your-business/environmental-impact/waste/producer-responsibility/.

3.2.3 Waste to Energy

The Netherlands has adopted several waste-to-energy tactics to manage its waste sustainably. Incineration remains the most common method of waste-to-energy conversion in the country. Municipal solid waste is incinerated in large-scale facilities, and the heat is used to generate electricity and provide heating for homes and businesses [20]. Biogas production is another waste-to-energy tactic used in the Netherlands. Anaerobic digestion is used to break down organic waste, which is then used to generate electricity, heat, and transportation fuel. When the landfill sites are excluded, the produced biogas originates from biowaste/municipal waste, co-digestion of agricultural/municipal/industrial side streams, and sewage sludge [50]. Co-firing, which is the practice of burning both fossil fuels and renewable fuels (such as biomass) together to generate electricity, is also practiced in the Netherlands. The SDE+ scheme[5] provides grants for producing energy from biomass. Several coal-fired power plants in the country have been retrofitted to enable co-firing biomass, thereby reducing greenhouse gas emissions [16, 47]). Plasma gasification is a more recent waste-to-energy technology that uses extremely high temperatures to convert waste into a gas, that can then be used for electricity generation. The Netherlands has a number of plasma gasification facilities in operation [17].

While waste-to-energy tactics have played an important role in the Netherlands' waste management strategy, waste reduction and recycling remain the preferred options.

3.3 Operational Activities: Experiments

Operational activities, experiments, and actions typically have a short-term horizon and are often carried out in the context of innovation projects and programs, in business and industry, in politics, or civil society [23]. Actions at this level are often driven by individual ambitions, entrepreneurial skills, or promising innovations. Below are some examples of these innovative actions linked to their broader strategic context:

3.3.1 Reusing Produced Material

Reusing products and materials is a key aspect of a circular economy, where resources are kept in use for as long as possible [13]. This strategy involves finding new ways to use existing products and materials, either by repairing, refurbishing, or repurposing them. Promoting reuse, helps to reduce waste, conserve resources, and minimize the

---

[5] SDE stands for Stimulation of Sustainable Energy Production and Climate Transition (Stimulering Duurzame Energieproductie en klimaattransitie). With the SDE subsidy, the government stimulates the production of sustainable energy.

environmental impact of production and consumption [43]. There is growing interest in the potential of reuse as a strategy for achieving a more sustainable and circular economy, and many initiatives are being implemented to support reuse at both local and national levels [19, 44]. The Dutch government has implemented several actions and initiatives to encourage the reuse of products and materials as part of its waste management strategy. Here are a few examples:

- **Stimulating circular design:** Several initiatives have been launched to promote circular design and production, which involves designing products for durability, repairability, and reuse. For example, the "Circular Design Program" aims to facilitate collaboration between designers, producers, and other stakeholders to create circular products and services [27].
- **Second-hand marketplaces:** The government has also supported the development of online platforms for second-hand goods, such as Marktplaats and Vinted. These platforms allow individuals to buy and sell used items, reducing the amount of waste generated and encouraging reuse.
- **Repair cafes:** The Dutch government has also supported the development of "repair cafes", which are community spaces where people can bring their broken items to be fixed by volunteers. These cafes promote repair and reuse, as well as community building and knowledge sharing [26].
- **Circular economy hubs:** The Netherlands hosts several circular economy hubs, that foster collaboration between businesses and entrepreneurs in recycling and upcycling projects. One example is the BlueCity Lab in Rotterdam, which is an "Urban Living Lab" for circular economy experiments. Urban Living Labs (ULL) are spaces that enable experimentation with sustainability solutions, allowing urban actors to design, test and learn from socio-technical innovations [48]. BlueCity lab defines its mission to accelerate transition from linear to circular economy. The lab is an incubator specifically aimed at biocircular design and biobased technology, and it offers space to entrepreneurs who are turning waste into a valuable raw material. Participants includes biotechnologists, designers, and innovative companies, who carry out experiments with biomaterials, pursuing ideas like making leather from plants, and self-healing concrete from bacteria.

3.3.2 Recycling and Upcycling:

Recycling and upcycling are complementary waste management approaches aimed at reducing the amount of waste that ends up in landfills or in the environment. Recycling involves converting waste materials into new products, while upcycling focuses on transforming waste into higher-value products. In these experiments, various actors, including waste management companies, governments, consumers, and environmental organizations, work together to develop and implement innovative waste practices. There is a growing number of experiments and niches related to recycling and upcycling in the Netherlands, including:

- **Textile recycling:** Refers to the process of reusing or repurposing textile waste, such as clothing, carpets, and fabric scraps, to create new products. Textile recycling reduces landfill wastes, and the environmental impact of textile production. There are several initiatives in the Netherlands focused on recycling and upcycling textiles, as it represents one of the major sources of waste. One example is the Texperium, a knowledge and innovation center working toward a circular textile industry by developing new recycling and upcycling technologies. Their mission is to increase the percentage of recycled content used in the Dutch Textile industry [21].
- **Plastic upcycling:** Plastic waste is also a major environmental concern in the Netherlands, and the country has invested in recycling and upcycling facilities to transform plastic waste into new products. For example, plastic waste is being recycled into new plastic products, and old clothes made from synthetic fibers like polyester (a form of plastic), are upcycled into new fashion items. There are several citizen-driven initiatives such as Precious Plastic, operating within the open loops of companies that have produced products and/or packaging. These initiatives consist of individual or interconnected, like-minded, local sustainability groups, with nearly 300 workspaces around the world where people recycle plastics with 'homemade' plastic recycling machines [42].
- **Food waste:** The Netherlands has several initiatives focused on reducing food waste, such as PeelPioneers, the largest Processing facility in Europe for orange peels derived from fresh orange juice production in supermarkets, hotels, and restaurants. These peels are transformed into a range of products suitable for the food industry, including dietary fibers for meat substitutes, bakery products, and sauces, thereby contributing to the shift towards protein alternatives. Additionally, the facility generates orange oil and other valuable raw materials that are in high demand in the food additives, detergents, and cosmetics industries [28].
- **E-Waste:** In the Netherlands, wasted electric and electronic equipment (e-waste) is a key topic on the circular economy agenda. This type of waste can be recycled either at a minimum standard typically through shredding, or at a higher standard that must meet WEEELABEX requirements,[6] to minimize hazardous substance emissions and promote material recovery. High-standard recycling is mandatory in the Netherlands. The non-profit organization "Wecycle" manages the structured collection and recycling of e-waste on behalf of 1,500 producers and importers, working in partnership with municipalities, shops, and installation companies.[7] Wecycle, active since 1999, was one of the first organizations in Europe to manage e-waste. In addition, Wecycle is responsible for the safe removal and destruction of (hydro) chlorofluorocarbons found in cooling and freezing appliances, which are harmful to the ozone layer and contribute to global warming [14].

---

[6] WEEELABEX Standards, http://www.weeelabex.org/standards/.

[7] Wecycle, https://www.wecycle.nl/.

To conclude, experiments and niche innovations in circular waste management systems are essential for driving the transition toward a more sustainable and circular economy. These experiments offer opportunities for testing and refining new ideas, technologies, and practices, and can ultimately lead to systemic change and transformation. However, the success of these experiments depends on a supportive ecosystem that enables experimentation, learning, and collaboration across different sectors and stakeholders. By fostering innovation and establishing such an enabling environment, the Netherlands and other countries can pave the way for a more sustainable and circular future.

3.4 Reflexive Activities: Monitoring and Evaluation

Reflexive activities include a range of processes such as monitoring, assessment, and evaluation of policies and societal change. These activities are not limited to formal institutional settings but are also socially embedded, with media and the Internet playing an influential role in shaping public opinion and assessing the effectiveness of political agendas. Scientific research also plays a crucial role in identifying longer-term societal trends and dynamics, and bringing them to the attention of policy-makers. Reflexive activities are vital for preventing systemic lock-ins and promoting the exploration of new ideas and transition trajectories. Monitoring processes and reflexive activities are applied to all categories discussed above (strategic, tactical, and operational activities) [23].

3.4.1 Organizations

Waste management plans and activities in the Netherlands are monitored by various governmental and non-governmental organizations to ensure that waste is handled in a safe, sustainable, and environmentally responsible way. Some of these organizations that work on assessing and monitoring waste management activities and aim to find bottlenecks and improve the process, include:

The Ministry of Infrastructure and Water Management is responsible for developing policies related to waste management in the Netherlands [29]. It sets the legal framework for waste management, including regulations related to waste reduction, reuse, and recycling. The policies are implemented at the regional level by Regional Waste Management Companies (RWMCs), which are often private or semi-public entities operating under municipal or inter-municipal oversight. RWMCs are responsible for collecting and processing waste in specific regions of the Netherlands. They are required to meet strict environmental standards, and are monitored by the government to ensure compliance [15].

Planbureau voor de Leefomgeving (PBL) or (Netherlands Environmental Assessment Agency) is an independent research institute that focuses on the environment, nature, and spatial planning. It was established in 2008 when the Netherlands

Institute for Spatial Research (Ruimtelijk PlanBureau-RPB) merged with the Netherlands Environmental Assessment Agency (Milieu- en Natuur PlanBureau-MNP). PBL operates under the Dutch government, specifically the Ministry of Infrastructure and Water Management, but may also carry out research at the requests of other government departments such as the Ministry of Economic Affairs and Climate Policy, the Ministry of the Interior and Kingdom Relations, the Ministry of Agriculture, Nature and Food Quality, and the Ministry of Foreign Affairs. PBL's main responsibilities are as follows:

1. Monitoring and reporting on the current state of environmental, ecological, and spatial quality, and assessing policies related to these areas.
2. Analyzing potential social trends that may affect environmental, ecological, and spatial quality in the future, and evaluating different policy options accordingly.
3. Identifying and highlighting key social issues that impact environmental, ecological, and spatial quality.
4. Identifying potential strategic approaches for achieving government goals in the areas of the environment, nature, and spatial planning [31].

The Dutch Association of Waste Management Companies (Vereniging Afvalbedrijven-VA) and Stichting Afvalfonds are private organizations that represent the interests of waste management companies in the Netherlands. They work closely with the government to develop and implement waste management policies. The Waste Management Association plays a connecting role in the transition to a circular economy. With more than fifty members, it represents two-thirds of the waste market, both in terms of waste volume and market share, in the entire Dutch waste chain. The members work on a wide range of activities, including prevention, collection, transport, sorting, cleaning, processing, recycling, composting, fermentation, sewage management, incineration, and landfill. The association has defined its mission as promoting the transition to the circular economy, with a focus on closing the cycle, a recovery of materials, raw materials, and energy [46].

3.4.2 Programs and Tools

Waste policy in the Netherlands is monitored by collecting, analyzing, and reporting on data at the municipal and national levels. For example, the Dutch Ministry of Infrastructure and Water Management's Circular Economy Program, monitors progress towards circularity goals by tracking indicators such as the percentage of recycled materials in products, the amount of waste generated per capita, and the number of circular procurement contracts. Much of the data is available through an online waste database [37]. The ministry also publishes regular reports on the circular economy (some of them are listed in Table 1).

The full list of reports can be found on the website of the Ministry of Infrastructure and Water Management. Below, some of the main programs and tools used for the assessment and monitoring system transition are briefly introduced:

Table 1 Waste circular-knowledge center circular economy [39]

Report date	Topic	Report name
Waste figures		
2023–03	Waste processing in the Netherlands, data 2021	Afvalverwerking in Nederland, gegevens 2021
2023–02	Waste tax 2022	Afvalstoffenheffing 2022
2022–05	Composition of residual household waste, sorting analyzes 2021	Samenstelling huishoudelijk restafval, sorteeranalyses 2021
2020–09	Dutch Waste in Figs. 2006–2016	Nederlands Afval in Cijfers 2006–2016
Circular economy		
2023–02	Rijkswaterstaat reuse strategy advice: reuse potential & action perspective	Advies strategie hergebruik Rijkswaterstaat: hergebruikpotentie & handelingsperspectief
2022–06	Evaluation of the ban on free plastic bags 2020 and 2021	Evaluatie verbod op gratis plastic tassen 2020 en 2021
2021–08	Application and assessment of Circular Materials Strategy	Rapportage Toepassing en Toetsing Circulaire Materialenstrategie
2020–02	The future starts now: annual report 2019: Rijkswaterstaat Impulse Program Circular Economy	De toekomst begint nu: jaarrapportage 2019: Impulsprogramma Circulaire Economic Rijkswaterstaat
2019–09	In-depth circular economy guide for MIRT projects	Verdiepende handreiking Circulaire Economie voor MIRT- projecten
2019–05	Engineering sustainability in the Netherlands	Rijkswaterstaat: proudly engineering sustainability in the Netherlands
2019–05	Inspiration book for an integrated approach to circular design	Inspiratieboek voor integrale aanpak circulair ontwerpen
2018–12	Circular design, added value in infrastructure	Circulair ontwerpen, meerwaarde in de infra
2017–12	Circular economy indicators	Circulaire economie indicatoren voor Rijkswaterstaat
2017–11	Implementation of raw materials efficiency	Eindrapportage implementatie grondstoffenefficiëntie

The Work Programme on Monitoring and Evaluation Circular Economy 2019–2023 is a collaboration, under the supervision of PBL (Netherlands Environmental Assessment Agency), involving a group of institutions.[8] It aims to monitor and

---

[8] The collaborators of this program are:
- Copernicus Institute of Sustainable Development (Utrecht University).
- CPB Netherlands Bureau for Economic Policy Analysis (CPB).
- Institute of Environmental Sciences (CML, Leiden University).
- National Institute for Public Health and the Environment (RIVM).
- Netherlands Enterprise Agency (RVO.nl).
- Netherlands Organisation for Applied Scientific Research (TNO).
- Rijkswaterstaat (Government Service for Roads and Waterways).

evaluate the progress towards the 'Netherlands Circular in 2050' plan, and to provide the knowledge base needed for informed policymaking process. The report resulting from this program (first integral circular economy report—ICER) was published in 2021. It presented the current state of the circular economy transition in the Netherlands, and described potential starting points and next steps to support further progress in this transition process [6].

The Circular Benchmark Tool (CBT) is a practical assessment instrument to understand, visualize, and compare the transition toward a circular economy across regions and provinces. This tool is licensed by ProActBlue and Rademaker (consulting under Attribution-NonCommercial- NoDerivatives 4.0 International). CBT serves as a replicable tool for knowledge exchange aimed at improving regional circular performance and facilitating cross-regional learning in the field of circular economy. CBT introduces 6 main areas of assessment including:

1. Circular procurement: The extent to which regions engage public and private suppliers to adopt circular economy practices in procurement.
2. Access to funding: The extent to which regions provide access to finance circular initiatives.
3. Circular Society: The extent to which regions raise awareness, motivate citizens, and develop circular competencies (knowledge and skills) to support circular economy transition.
4. Value Chain Activation: The extent to which the region fosters circular value creation among stakeholders within and across supply chains.
5. Good Governance: The extent to which the region coordinates, facilitates, promotes, and supports transition to a circular economy.
6. Integrated Policy Framework: The extent to which the region establishes and integrates a coherent circular policy framework over the regional policy domains, to align policy and legislation with circular principles and best practices.

Using this self-assessment tool, the regional authorities can evaluate, understand, and improve their Circular Maturity Level (including stages such as mapping, planning, doing, checking and acting, and leading by example). The regions can obtain reports with local results and benchmarks of other regions. They can also learn from a wide literature base of best practices and examples, and apply their regional experiences and findings as input for Circular Economy Action Plans (CEAP), thereby accelerating the circular economy transition through data-driven decision-making and quantitative tools [7].

- Statistics Netherlands (CBS).

4 Conclusion

This chapter provides a comprehensive overview of the process and experiences of managing the transition to a circular waste management system in the Netherlands. The transition management process introduces a cycle that involves four categories of activities (strategic, tactical, operational, and reflexive activities). The first step is setting the transition arena where problems are identified, and goals and objectives are defined by engaging different groups of stakeholders and developing a shared vision among them. In the Netherlands, the key problems were the high levels of waste generation, limited recycling and reuse, and environmental degradation, identified over a period of several decades starting around the 1980s. In response the government decided to reduce the amount of landfilled waste by reducing waste generation, recovering valuable raw materials from waste, generating energy by incinerating residual waste, and landfilling only what's left in an environmentally sound way. The national agreement on circular economy (NACE) introduced set the long-term vision of becoming fully circular by 2050, and introduced the policies and plans to achieve this ambitious goal. In the next step for defining a transition agenda, five sectors and value chains were introduced as the focus areas, because they had a high environmental impact. These areas were: plastics, consumer goods, manufacturing, construction, and biomass and food. These sectors provide the foundation for implementing tactical activities such as waste prevention, waste separation, and waste-to-energy activities. Operational activities introduce action plans that focus on reusing produced materials, through circular design, promoting second-hand marketplaces, repair cafés, and circular economy hubs. The next group of operational activities focuses on recycling and upcycling, recycling, and upcycling textiles, plastic, food, and E-waste. The final category introduces the reflexive activities, including the assessment and monitoring the transition process to ensure progress towards the desired goals, and to adjust the strategies and activities the least performing areas. The organizations like ministry of infrastructure and water management, the Netherlands Environmental Assessment Agency, and the Dutch association of waste management companies are active in defining the indicators and measuring system performance.

Real-life examples from the Dutch experience of circular waste management systems show that this approach can be successful, and organizations can achieve a range of benefits, including reduced waste, increased resource efficiency, and improved environmental performance. However, the transition to a circular waste management system is a complex and ongoing process that requires ongoing commitment and collaboration from all stakeholders.

Overall, the successful transition to a circular waste management system requires a strategic, integrated approach that addresses the entire waste management system, from waste generation to final disposal. By implementing strategic, tactical, operational, and reflexive activities, organizations can effectively manage the transition to a circular waste management system and make a significant contribution to a more sustainable future.

References

1. Avelino F, Wittmayer JM, Kemp R, Haxeltine A (2017) Game-changers and transformative social innovation. Ecol Soc 22(4)
2. Bäckstrand K, Lövbrand E (2019) The road to Paris: contending climate governance discourses in the post-Copenhagen era. J Environ Policy Plan 21(5):519–532
3. Berkhout F, Stirling A, Smith A (2004) Socio-technological regimes and transition contexts. Syst Innov Trans Sustain: Theory Evid Policy 44(106):48–75. https://doi.org/10.4337/9781845423421.00013
4. Bhada-Tata P, Hoornweg DA (2012) What a waste?: a global review of solid waste management. World Bank Group. United States of America. https://doi.org/20.500.12592/dk0mgg. Retrieved from: https://openknowledge.worldbank.org/entities/publication/1a464650-9d7a-58bb-b0ea-33ac4cd1f73c,
5. Bocken NM, De Pauw I, Bakker C, Van Der Grinten B (2016) Product design and business model strategies for a circular economy. J Prod Eng 33(5):308–320
6. Brink H, Koch J, Prins AG, Rood T, Hanemaaijer A, Kishna M, Schoenaker N (2021) Integrale Circulaire Economie Rapportage 2021. PBL Planbureau voor de Leefomgeving
7. CBT (2023) Circular benchmark tool. ProActBlue and rademaker Consulting
8. Dimitropoulos A, Tijm J, In 't Veld D (2021) Extended producer responsibility: design, functioning and effects. CPB Netherlands Bureau for Econ Policy Anal PBL Netherlands Environ Assess Agency. https://www.cpb.nl/sites/default/files/omnidownload/PBL-CPB-2021-Extended-Producer-Responsibility-Design-Functioning-Effects.pdf
9. Ellen MacArthur Foundation (2015) Delivering the circular economy: a toolkit for policymakers. Ellen MacArthur Foundation
10. Eurostat (2021) Municipal waste statistics. https://ec.europa.eu/eurostat/statistics-explained/index.php?title=Municipal_waste_statistics
11. Eurostat (2023) Recycling rate of municipal waste. https://ec.europa.eu/eurostat/databrowser/view/ten00063/default/table
12. Frantzeskaki N, De Haan H (2009) Transitions: two steps from theory to policy. Futures 41(9):593–606. https://doi.org/10.1016/j.futures.2009.04.009
13. Ghisellini P, Cialani C, Ulgiati S (2016) A review on circular economy: the expected transition to a balanced interplay of environmental and economic systems. J Clean Prod 114:11–32
14. Golsteijn L, Valencia Martinez E (2017) The circular economy of E-waste in the Netherlands: optimizing material recycling and energy recovery. J Eng 2017(1):8984013
15. Goorhuis M, Reus P, Nieuwenhuis E, Spanbroek N, Sol M, van Rijn J (2012) New developments in waste management in the Netherlands. Waste Manag Res 30(9):67–77
16. Government of the Netherlands (n.d.) Circular Dutch economy by 2050. Retrieved July 5, 2023, from https://www.government.nl/topics/circular-dutch-economy-by-2050
17. Hrbek J (2016) Status report on thermal biomass gasification in countries participating in IEA Bioenergy Task 33. IEA Bioenergy. Paris, France
18. Kemp R, Schot J, Hoogma R, (1998) Regime shifts to sustainability through processes of niche formation: the approach of strategic niche management. Technol Anal Strateg Manag 10(2):175–198
19. Kirchherr J, Reike D, Hekkert M (2017) Conceptualizing the circular economy: an analysis of 114 definitions. In Resources, conservation and recycling, vol 127. https://doi.org/10.1016/j.resconrec.2017.09.005
20. Milios L, Reichel A (2013) Municipal waste management in the Netherlands, European Environmental Agency
21. Looman A (2019) Building towards Circularity, the role of Business collaboration in the transition to-wards a circular Dutch textile industry
22. Loorbach D (2007) Transition management. New mode of governance for sustainable development. International Books
23. Loorbach D (2010) Transition management for sustainable development: a prescriptive, complexity-based governance framework. Governance 23(1):161–183

24. Markard J, Raven R, Truffer B (2012) Sustainability transitions: an emerging field of research and its prospects. Res Policy 41(6):955–967
25. Meadowcroft J (2009) What about the politics? Sustainable development, transition management, and long term energy transitions. Policy Sci 42(4):323–340. https://doi.org/10.1007/s11077-009-9097-z
26. Moalem RM, Mosgaard MA (2021) A critical review of the role of repair cafés in a sustainable circular transition. Sustainability 13(22):12351
27. NACE (Netherlands Circular Economy Programme) (2016) A circular economy in the Netherlands by 2050: government-wide programme for a circular economy 2019–2023. Ministry of Infrastructure and Water Management. Retrieved from https://www.government.nl/documents/reports/2016/09/14/a-circular-economy-in-the-netherlands-by-2050
28. Neofotistos M, Hanioti N, Kefalonitou E, Perouli AZ, Vorgias KE (2023) A real-world scenario of citizens' motivation and engagement in urban waste management through a mobile application and smart city technology. Circ Econ and Sustain 3(1):221–239
29. Nijboer NMA (2021) The changing process of public procurement through the integration of circularity in a public organization: a challenge or an opportunity? (Master's thesis). Utrecht University
30. OECD (2011) Towards green growth: monitoring progress–OECD indicators. OECD Publishing. https://doi.org/10.1787/9789264111318-en
31. PBL Netherlands Environmental Assessment Agency (n.d.) Netherlands environmental assessment agency. Retrieved January 31, 2023, from https://www.pbl.nl/en
32. Pielke R (2018) Opening up the climate policy envelope. Issues Sci Technol 34(4):30–36
33. Rijksoverheid (n.d.) Huishoudelijk afval scheiden en recyclen. Retrieved November 20, 2022, from https://www.rijksoverheid.nl/onderwerpen/afval/huishoudelijk-afval-scheiden-en-recyclen
34. Rijksoverheid (2016) National waste management plan 2017–2029. Retrieved December 12, 2022, from https://www.rijksoverheid.nl/documenten/beleidsnota-s/2016/12/21/landelijk-afvalbeheerplan-2017-2029
35. Rijksoverheid (2018) Accelerating the transition to a circular economy. Retrieved January 15, 2023, from https://www.rijksoverheid.nl/documenten/rapporten/2018/02/15/versnelling-naar-een-circulaire-economie
36. Rijkswaterstaat (n.d.-a) Landelijk afvalbeheerplan. Afval circulair—kenniscentrum circulaire economie. Retrieved March 17, 2023, from https://www.afvalcirculair.nl/onderwerpen/beleid-circulaire/landelijk/
37. Rijkswaterstaat (n.d.-b) Waste monitor database. Retrieved May 10, 2023, from https://afvalmonitor.databank.nl/
38. Rijkswaterstaat (2016) Government-wide program "The Netherlands circular in 2050". Retrieved May 2, 2023, from https://www.government.nl/topics/circular-economy/circular-dutch-economy-by-2050
39. Rijkswaterstaat (2023) Waste circular—knowledge center circular economy. Retrieved January 12, 2023, from https://www.afvalcirculair.nl/onderwerpen/linkportaal/publicaties/
40. Rijkswaterstaat Environment (n.d.) Elements of dutch waste management system. Retrieved January 21, 2023, from https://rwsenvironment.eu/subjects/from-waste-resources/elements-dutch-waste/
41. Rotmans J, Kemp R, Van Asselt M (2001) More evolution than revolution: transition management in public policy. Foresight 3(1):15–31. https://doi.org/10.1108/14636680110803003
42. Spekkink W, Rödl M, Charter M (2022) Repair cafés and precious plastic as translocal networks for the circular economy. J Clean Prod 380:135125
43. Stahel WR (2016) The circular economy. Nature 531(7595):435–438. https://doi.org/10.1038/531435a
44. Tukker A (2015) Product services for a resource-efficient and circular economy–a review. J Clean Prod 97:76–91

45. Turnheim B, Geels FW (2012) Regime destabilisation as the flipside of energy transitions: lessons from the history of the British coal industry (1913–1997). Energy Policy 50:35-49. https://doi.org/10.1016/j.enpol.2012.04.060
46. Vereniging Afvalbedrijven (n.d.) Vereniging Afvalbedrijven (Dutch Waste Management Association). Retrieved January 14, 2023, from https://www.verenigingafvalbedrijven.nl/eng
47. Van der Stelt MJC, Gerhauser H, Kiel JH, Ptasinski KJ (2011) Biomass upgrading by torrefaction for the production of biofuels: a review. Biomass Bioenerg 35(9):3748–3762
48. Von Wirth T, Fuenfschilling L, Frantzeskaki N, Coenen L (2019) Impacts of urban living labs on sustainability transitions: mechanisms and strategies for systemic change through experimentation. Eur Plan Stud 27(2):229–257
49. Westley F, Olsson P, Folke C, Homer-Dixon T, Vredenburg H, Loorbach D, Van Der Leeuw S (2011) Tipping toward sustainability: emerging pathways of transformation. Ambio 40(7):762–780
50. Winquist E. Van Galen M, Zielonka S, Rikkonen P, Oudendag D, Zhou L, Greijdanus A (2021) Expert views on the future development of biogas business branch in Germany, The Netherlands, and Finland until 2030. Sustainability 13(3):1148
51. Wittmayer JM, Schäpke N (2014) Action, research and participation: roles of researchers in sustainability transitions. Sustain Sci 9(4):483–496

Open Access This chapter is licensed under the terms of the Creative Commons Attribution 4.0 International License (http://creativecommons.org/licenses/by/4.0/), which permits use, sharing, adaptation, distribution and reproduction in any medium or format, as long as you give appropriate credit to the original author(s) and the source, provide a link to the Creative Commons license and indicate if changes were made.

The images or other third party material in this chapter are included in the chapter's Creative Commons license, unless indicated otherwise in a credit line to the material. If material is not included in the chapter's Creative Commons license and your intended use is not permitted by statutory regulation or exceeds the permitted use, you will need to obtain permission directly from the copyright holder.

Accumulation of Circular Economy Policy in China: Goals, Instruments and Demonstration Projects Over 2006–2022

Wenting Ma, Martin de Jong, and Thomas Hoppe

Abstract This chapter presents the accumulation of policies regarding Circular Economy (CE) in China over the 2006–2022 period, and focuses more particularly on developments in terms of policy goals, policy instruments and demonstration projects. Results show that the central government issued and implemented a growing number of CE policies and instruments over time, and more intensively following the 13th Five-Year Plan period (2016–2020), when a sudden increase was witnessed in terms of diversity of policy instruments, whilst demonstration programs moved from "CE Pilots" via "National Circular Economy Demonstration Cities" to "Zero-Waste Cities". Approximately simultaneously, governance and implementation of CE policies witnessed a shift from a primary focus on production efficiency via lowering of consumption patterns, to embracing a whole life cycle perspective, with policies covering the entire supply chain. The main reasons for this shift concern: (1) the government acknowledging that China's economic development had reached a stage which called for more attention to environmental protection; (2) CE becoming firmly tied to the national planning and strategy on "Ecological Civilization" at a time when there was a fierce competition between national government institutes; and (3) the

---

W. Ma (✉)
School of Economics and Management, Harbin Institute of Technology (Shenzhen), Shenzhen, China
e-mail: mawenting@hit.edu.cn

W. Ma · M. de Jong
Rotterdam School of Management & Erasmus School of Law, Erasmus University Rotterdam, Rotterdam, The Netherlands
e-mail: w.m.jong@law.eur.nl

M. de Jong
Institute for Global Public Policy, Fudan University, Shanghai, China

M. de Jong · T. Hoppe
Smart City Institute, HEC Liège, Université de Liège, Liège, Belgium
e-mail: t.hoppe@utwente.nl

T. Hoppe
Section of Governance and Technology for Sustainability (CSTM), Department of Technology, Policy and Society, Faculty of Behavioural, Management and Social Sciences, University of Twente, Enschede, The Netherlands

© The Author(s) 2026
M. de Jong et al. (eds.), *The Inclusive Circular Economy*, Urban Sustainability, https://doi.org/10.1007/978-981-96-6867-0_6

occurrence of policy learning from foreign CE programs, policies, demonstration projects, and practices.

Keywords Circular economy · Zero-waste pilots · Policy goals · Policy instruments · Demonstration projects · Policy accumulation

1 Introduction

Chinese cities have experienced rapid economic development and urban population expansion since the 1980s. The percentage of urban population nationwide has increased from less than 20% in 1978 to 64.7% in 2021 [27]. Along with this development economic prosperity and material quality of life have increased considerably. However, this rapid urban development has come at a price, with urban expansion causing a variety of social and environmental problems such as traffic congestion, air pollution, and poor living conditions [3, 44]. To respond to these challenges, the Chinese central government proposed several sustainable development visions, such as the "Ecological Civilization[1]" and "Beautiful China" [18]. In line with this, plenty of sustainable development policies were proposed—by both the national as well as the provincial and local governments—to balance environmental protection and economic growth. The central government also introduced several types of demonstration and pilot programs to promote the local implementation of these visions, such as the "eco city", "low carbon city", and "sponge city" [5, 18]. Among these, Circular Economy (CE) pilots can be considered an important component aimed at transforming current economic development strategies, particularly regarding urban waste problems. In China, CE can be viewed as a holistic approach through which policymakers seek to align environmental and economic development strategies.

The term "Circular Economy" was introduced in China during the 1990s, and originated in the fields of cleaner production, industrial ecology and ecological modernization. It was formally accepted in 2002 by China's central government as a new development pathway [8]. CE is widely perceived as an alternative model of production and consumption, a strategy which theoretically contributes to both economic growth and sustainable development whilst reducing environmental pollution. Key to CE strategy is the '3R' principle, focusing on reducing raw material and energy inputs, reusing materials, and recycle waste materials [33]. In China, CE is considered important by the central government as it contributes to national programs on ecological civilization and environmental quality. This is showcased by the sheer number of policies, CE practices, and demonstration pilots that have been presented by China's national government in recent years. In 2008, China's government issued

---

[1] Ecological civilization is the summary of material, spiritual and institutional achievements made by mankind to protect and build a beautiful ecological environment. It is a systematic project that involves all aspects of economic, political, cultural and social development. It reflects the progress of a society's civilization [12].

the Circular Economy Promotion Law, which was revised in 2018. This law has an important impact on CE policy evolution [21]. More particularly, the National Development and Reform Commission (NDRC) proposed the "Circular Economy Pilots" program in 2005. Subsequently, in 2013 and 2015, the NDRC gradually selected 102 cities and counties as "National Circular Economy Demonstration Cities" [31]. By implementing a great many CE policies and practices, the Chinese government argued that it had succeeded in greatly improving the quality of the domestic environment. However, waste disposal amounts were still considered astronomical in 2021 [2]. And whereas the NDRC stressed improvement of resource efficiency, the Ministry of Ecology and Environment emphasized environmental protection. In 2019, the State Council issued the concept of the "Zero-Waste City" (ZWC) to implement a comprehensive management reform for solid waste handling at the city level [38]. Here, zero waste refers to the conservation of all resources by means of responsible production, consumption, reuse, and recovery of products, packaging, and materials without burning and with no discharges to land, water, or air that threaten the environment or human health [46]. In 2019 the Ministry of Ecological Environment (MEE) selected 16 cities followed by 113 cities and counties in 2022 to become ZWC demonstration pilots.

Although China has managed to introduce CE in terms of increasing numbers of programs, policies and demonstration pilots, the systematic research into CE is still surprisingly limited, particularly when it comes to CE policy accumulation over time and the driving factors behind their sheer increase. To better understand the accumulation of CE policies over time, this chapter seeks to answer the following two research questions: (1) *How was CE implemented in China in terms of policy goals, instruments, and demonstration projects over 2006–2022?* and (2) *What factors influenced the CE policy accumulation process?*

This chapter is structured as follows. Section 2 reviews key literature on CE policy goals, policy instruments, and demonstration projects. Section 3 presents the changes in environmental governance concepts and visions in China. Sections 4, 5 and 6 present the accumulation of CE initiatives in terms of policy goals, instruments, pilot projects and key events from 2006 to 2022. Section 7 discusses and compares the accumulation of CE policies in the different five-year plans. Finally, Sect. 8 presents the conclusions of this study and offers suggestions for future research.

2 Introduction to CE Policy

To improve one's understanding of CE policy accumulation, an overview is given of state-of-the art literature related to CE policies in China. Policy evolution is the phenomenon of adjusting public policies in response to evolving social, political, and economic circumstances. On the other hand, policy accumulation emerges when policy makers introduce new regulations without eliminating existing ones [16]. New policies may coexist with old ones without entirely replacing or erasing them (Pollitt and Bouckaert 2011). In this section we will introduce the concept of policy goals and the tools that are used to implement them, policy instruments.

2.1 CE Policy Goals

A policy typically has one or more goals. Policy goals can be seen as the intended effects of policy actions taken by relevant policy implementing entities in order to solve or mitigate relevant social and public issues [11]. Some concepts are similar to policy goals, such as objectives and targets. Whereas the former indicates the direction of an intended action but not the action itself [13], the latter is usually narrower, more specific, and defines the thematic field in which policy action is planned [39]. In comparison, policy goals are considered to be at a higher level of abstraction than policy objectives and targets [17]. More in general, in a normative-prescriptive sense, a policy goal should be clear, achievable, future-oriented, and coordinated in the implementation process. For example, policy goals of the European Union's CE Policy Strategy include: reduction in material inputs, and increase in gross domestic product (GDP) due to material cost-saving opportunities [6]. Morseletto [24] proposed that the goals of CE can be divided into five main areas, including efficiency, recycle, recovery, reduction and design. A policy goal often has a national connotation and reflects the effects or impact of a grand strategy for the entire country. Policy instruments, then, can be seen as concrete tools used to achieve policy goals throughout policy implementation and related practices. Finally, it should be noted that policy goals influence the selection of policy instruments [40].

2.2 CE Policy Instruments

When reviewing the CE policy literature, multiple policy instruments are found that have the potential to advance or reach certain CE goals. Policy instruments can be seen as the building blocks in policy implementation processes [35]. CE policy instruments are implemented to promote resource-efficient and sustainable use of natural resources thus contributing to the CE. CE policy instruments can vary in nature and may include bans of hazardous substances, taxes on use of a natural resource, or extended producer responsibility [14].

Presently, four types of CE policy instruments are discerned to capture potential differences in the availability and use of government resources and incentive mechanisms, including legal policy instruments, economic policy instruments, network policy instruments, and communicative policy instruments. This classification of policy instruments is also used in other domains of environmental policy, such as low carbon city development [9, 10, 19].

Closely related to the concept 'policy instrument' is the 'policy mix'. The latter includes one or more specific goals or objectives of a certain policy field, as well as a combination of (different types of) policy instruments [26]. Individually, each policy instrument offers certain benefits and limitations, yet collectively policy mixes offer the possibility to provide synergies and enhance effectiveness for achieving integrated CE goals and outcomes, that exceed the impact of policy instruments individually. Policy mixes typically contain a combination of multiple policy instrument types, with the aspiration that different types reinforce each other. Either within a policy mix or not, policy instruments interact with each other due to their different functions [15, 34].

When the number and diversity of policies increase over a given period policy accumulation occurs. Policy accumulation can be considered as a process of continuous expansion of public policies [1]. It involves the growth in the number of policies, enriching the types of policy goals, and increasing the number and variation in types of policy instruments [20]. However, it may also cause increased complexity in policy packages [16]. Zhu et al. [45] discussed China's CE policy effort and analyzed the policies proposed by the central government and various of its ministries. Drivers for the CE policy accumulation include international policy learning and striking a balance between economic and environmental pressures [20, 45].

3 CE Demonstration Programs in China

In China, the central government selects a few demonstration pilots when it wishes to introduce new policies and checks what consequences their implementation will have (as for CE). Pilot copying and popularization is an important method for China's authorities to govern the country and realize reforms. When demonstration pilots are considered successful in achieving their goals, the central government decides to transfer successful local experiences, and replicate them elsewhere in the country. Pilot programs may include pilot cities, pilot projects and pilot measures, and they are typically related to certain frames, philosophies, slogans, or narratives. For example, in 2007, a new philosophy entitled "Ecological Civilization" was launched to advocate harmonious coexistence between nature, individuals, and society at large [4]. It is a slogan for China's sustainable development, and it was used to promote industrial transformation. In line with this a series of sustainable development pilot projects were proposed by China's State Council and ministerial departments, ranging from the Low-Carbon Eco City programs in 2010, to the CE pilots and the more recent ZWC pilot programs in 2019. More than 250 cities in China have introduced either eco or green pilot projects [42]. Cities are major contributors to environmental problems and resource consumption. Urbanization is expected to increase further in the next decades. The implementation of CE in cities will continue to be a crucial part in solving resource constraints and environmental problems [41]. CE pilots can be viewed as the basic pathway for developing China's ecological civilization. In addition, pilot programs may include demonstration pilot schemes emphasizing certain policy instruments, like pilots addressing carbon trading [25]. Some scholars have studied the effectiveness of pilot schemes (e.g., emission trading) in environmental governance [7]. Pilot programs can also be seen as new attempts to explore economic reform and development in a selected region or place, such as Pudong New Area, Shenzhen Special Economic Zone, or Xiong'an New Area.

4 Changes into Environmental Governance Visions (2003–2022)

After the opening up and reform in 1978, the pattern of economic development in China shifted towards a market-oriented, fast-paced growth paradigm. To promote economic development, "pollute first, control later" was the main pattern of urban development in this period. Environmental governance was considered inferior and easily neglected in modern urban expansion processes. However, with the deterioration of the natural environment, ecological governance received more attention from policymakers. The idea to develop a circular economy was proposed by China's State Council in 2005, when the national government adopted a "Resource-conserving and environment-friendly" logic [36]. In 2005, the State Council issued several *Opinions on Accelerating the Development of Circular Economy* as a starting symbol of China's circular economic development. To unify economic, environmental and social development, the central government proposed to adopt various measures to maximize economic output and minimize waste production with minimum resource consumption and environmental cost, whilst using the "reduction, reuse and resource recovery" principle [36]. At this time the national government also altered its development paradigm and adopted a more sustainable developmental pathway. It proposed a variety of solutions to respond to environmental pollution. For example, it proposed "Scientific Outlook on Development" in 2003 to promote harmonious development between humanity and nature. In 2006, at the Sixth National Conference on Environmental Protection the central government proposed to shift the emphasis toward a balance between environmental protection and economic growth. In 2007, a new philosophy entitled "Ecological Civilization" was launched to advocate the harmonious coexistence between nature, individuals, and society at large [4]. At the 18th National Congress of the CPC (2012), it was proposed to develop a circular economy and promote the reduction, reuse, and recycling of production, circulation, and consumption. At the same time, the position of CE was elevated to a new strategic level by making the establishment of a full-fledged resource recycling system one of the goals to build an all-round moderately prosperous society by 2020. In 2016, it was embraced in the 13th Five-Year socio-economic Plan (FYP). Subsequently another term—"Beautiful China"—was put forward as a key governing concept/frame at the 18th National Congress of the Communist Party of China (CPC). In 2017, the importance of developing an ecological civilization, and "Building a beautiful China", was once again mentioned in the 19th National Congress of the CPC [43]. The ecological slogan "Lucid waters and lush mountains are invaluable assets" was introduced and

Table 1 Overview of the key environmental governance strategies in China (2003–2022)

Year	Key conference	Key environmental governance strategies and thoughts	Guiding ideology/key development concept/ key words	Five-year planning period
2003		Scientific outlook on development	Scientific outlook on development	10th FYP
2007	17th National CPC Congress[2]	Concept 'ecological civilization' first appears at the 17th National Congress	Scientific outlook on development	11th FYP
2012	The 18th National Congress of CPC	'Ecological civilization' is adopted by the CPC charter at the 18th National Congress	Ecological civilization	12th FYP
2017	The 19th National Congress of CPC	• Accelerate institutional reform of ecological civilization progress and build a beautiful China • A new vision of innovative, coordinated, green, open and shared development • "Lucid waters and lush mountains" are invaluable assets (Proposed in 2005 and kicked off in 2017)	Harmonious coexistence between human and nature, beautiful China, ecological civilization	13th FYP
2020		China would aim to achieve peak CO_2 emissions before 2030 and carbon neutrality before 2060	Peak emissions and carbon neutrality	13th FYP
2022	The 20th National Congress of CPC	• Promote green development and harmonious coexistence between human and nature • Enhance the diversity, stability and sustainability of the ecosystem	• Green transformation • Environmental pollution prevention • Promote carbon peaking and carbon neutrality	14th FYP

used as a frame since 2017. During the National Ecological Environmental Protection Conference in 2018, President Xi Jinping coined environmental problems as a priority to the people's livelihood. In the same year the ecological civilization was adopted and codified in China's Constitution. In September 2020, China announced its target of being carbon–neutral before 2060 after reaching its peak in 2030. During the 20th National Congress of the CPC (2022), Beijing emphasized green development and green transformation, thus preventing environmental pollution, improving ecosystem diversity, and actively yet prudently promoting carbon peaking and carbon neutrality. Table 1 presents the key environmental governance concepts over the 2003 to 2022 period.

---

[2] CPC refers to Chinese Communist Party.

5 CE Policy Development and Implementation in 11th FYP (2006–2010)

5.1 CE Policy Goals in 11th FYP

In line with the *Outline of the 11th Five-Year Plan for National Economic and Social Development*, the central government proposed to develop and promote transformative changes to CE in a number of industries, including the steel, non-ferrous, coal, electric power, chemical, building materials, and sugar industries, whilst identifying a number of CE demonstration enterprises. In this period, the goal of the central government was to build a resource-saving and environment-friendly society. The basic state policy focused on conserving resources whilst protecting the environment, and building a circular and sustainable national economic system with low input, high output, low consumption and low emissions, and a resource-saving and environment-friendly society. To develop a CE, the government would attach equal importance to development and conservation, give priority to conservation using strategies relating to the principles of reduction, reuse and resource recovery. China would establish a resource recycling system for the whole society in the links of resource exploitation, production and consumption, waste generation and consumption.

5.2 CE Policy Instruments in 11th FYP

The number of CE policy instruments is shown in Table A1 (Appendix, using a classic classification of instruments). In the 11th FYP, the central government used legal instruments to eliminate backward production capacity. More economic instruments were adopted to support cleaner production and energy savings, such as subsides and dedicated and earmarked funding schemes. The functions of these economic instruments were to reduce emission and conserve the energy. Some commercial companies were involved in this period to save energy in the electricity industry. In this period, the goal was to reap primarily economic benefits, and consequently economic instruments were prioritized over legal policy instruments. Economic measures are easier to put into action and more readily accepted by companies, because it allowed them to improve their production efficiency and reduce their running costs.

5.3 NDRC's "Circular Economy Pilots" (11th FYP)

The "Circular Economy Pilots" program was introduced by the National Development and Reform Commission (NDRC). The construction cycle of the pilot circular economy ran from 2005 to 2010. In 2007, NDRC issued the *"Evaluation Index System for Circular Economy"*, which included 22 sub-indicators for CE, and could be divided into four categories: resource output, resource consumption, comprehensive utilization of resources, and waste discharge [29]. The NDRC published two batches of CE pilot lists in 2005 and 2007, respectively. In total, the "CE Pilots" program was mainly implemented in seven provinces and twenty cities. Except the pilot provinces and cities, the NDRC also paid more attention to circular development in some key industries (iron and steel, coal, electric power, etc.), fields (recycling of renewable resources, remanufacturing, etc.) and industrial parks [36]. The *Law of the People's Republic of China on the Promotion of Circular Economy* was adopted by the Standing Committee of the 11th National People's Congress of the People's Republic of China in 2008 and was implemented from 2009 on.

6 CE Policy Development and Implementation in 12th FYP (2011–2015)

6.1 CE Policy Goals in 12th FYP

In the 12th FYP, China's national government proposed to strengthen planning guidance and financial policy support, improve laws, regulations and standards, implement the extended producer responsibility system, drew up lists of circular economy technologies and products, established a labeling system for recycled products, and improved a statistical evaluation system on CE. China's national government also proposed to develop and apply technologies of resource reduction, recycling, remanufacturing, zero emission and industrial linkage, and promote demonstration models of CE. The concept of green development was also issued in the 12th FYP period. Here, government was to continue building a resource-efficient and environmental-friendly society. In addition, the central government was to actively respond to global climate change, whilst strengthening resource conservation and management, as well as vigorously developing a CE.

6.2 CE Policy Instruments in 12th FYP

In the 12th FYP, China's national government continued adopting more economic rather than legal policy instruments. More tax incentives and governmental grants were used to develop CE, such as dedicated funds for government procurement of energy-saving products, economic incentives for upgrading energy-saving technologies, and earmarked funds for circular economic development. More industrial standards and legal instruments were introduced in the same period. In 2011 the central government proposed to establish the legal system on CE. To be specific, the Ministry of Industry and Information Technology (MIIT) proposed to erect a technical standard system for energy saving and comprehensive utilization in industry and communication. More plans and documents were designed to develop CE, such as *Guidelines on Comprehensive Resource Utilization during the 12th FYP Period*, *The 12th FYP for Comprehensive Utilization of Bulk Industrial Solid Wastes*, "*Environmental protection" for Scientific and Technological Development in the 12th FYP*, *The Construction Plan of Ecological Civilization Pilot Demonstration*, *Strategic Action Plan of Energy Development*, etc. The number of network and communicative instruments increased steadily in this period. For example, more industry associations, other intermediaries, and experts were invited to the elaboration of CE policies. Some private organizations, such as catering companies, and individuals were encouraged to engage in green production and consumption patterns. The central government adopted more educational activities and exhorted people to adopt sustainable lifestyles.

In this period, central government issued several regulations regarding waste electrical and electronic products management regulations. It drew lessons on the EPR system (Extended Producer Responsibility) from Europe, and expanded it to the automotive, packaging, and other economic sectors.

6.3 NDRC's "National Circular Economy Demonstration City (County)" (12th FYP)

In 2013, the State Council issued a *Circular Economic Development Strategy and Near-Term Action Plan*. In 2015, the National Development and Reform Commission (NDRC) issued the *Circular Economy Promotion Plan 2015*. In 2013 and 2015, the NDRC gradually selected 102 cities and counties as "National Circular Economy Demonstration Cities (Counties)" [31]. At this stage, the "National Circular Economy Demonstration City (County)" program covered more small and medium-sized cities and county-level cities. The CE index system was revised in 2016. A series of plans and measures were adopted in this period. For example, in 2013, the State Council published the *CE Development Strategy and Near-term Action Plan*.

7 CE Policy Development and Implementation in 13th FYP (2016–2020)

7.1 CE Policy Goals in 13th FYP

In the 13th FYP, the national government proposed a plan to strengthen the connection between production and consumption and speed up waste recycling processes. The nation's industrial structure should be adjusted in line with the material flows and the use of recycled materials should be increased in comparison with raw materials. The number of circular parks was also to grow. A complex industrial and agricultural CE demonstration zone should be developed to facilitate the coupling and symbiosis between various enterprises within industrial park and across industries. The central government encouraged the development and utilization of urban mines, making good use of industrial solid waste and other bulk waste, speed up the development of urban kitchen waste, construction waste, textile waste and standardize the development of remanufacturing. It intensified efforts to improve the ecological environment and promote energy conservation and began to promote building a water-saving society. The central government also aimed to strengthen the conservation and management of mineral resources. Finally, to develop CE goals, the central government argued that society should adopt a thrifty lifestyle and improve the efficient use of resources.

7.2 CE Policy Instruments in 13th FYP

In the 13th FYP, the volume of CE policies (goals and instruments) grew to vast proportions. The number of policy instruments of all types increased substantially, particularly the economic and regulatory ones. In this period, the central government paid specific attention to responsibilities of producers. This led to the introduction of more legal instruments besides the economic ones. Environmental performance evaluation is closely linked with local government assessment and the promotion of local officers. Economic measures, such as government preferential policies, grants, and green loans were still commonly used in the 13th FYP. A mass of dedicated funds was given as input to reduce pollution volumes and diverse investment and funding channels were opened and stepped up to promote CE. During the 13th FYP, more network-based instruments were adopted as well. The central government actively stimulated inter-ministerial coordination in clean energy policies. The involvement of research institutions and private enterprises was deemed highly supportive for implementing the CE. A variety of media channels and platforms were created to disseminate knowledge regarding environmental protection. Citizens were to be persuaded to consume green products and enjoy green lifestyles.

7.3 MEE's "Zero-Waste City" (13th FYP and 14th FYP)

In 2015, NDRC proposed the *CE Promotion Plan*. In 2017, the NDRC issued the *Circular Development Leading Action*. The *Law of the People's Republic of China on Promoting the Circular Economy* was amended at the Sixth Session of the Standing Committee of the 13th National People's Congress in 2018.

Subsequently, the ZWC was issued by the State Council after the 19th National Congress of the CPC. In this period, the main governing idea of the central government was to "enhance ecological civilization and build a beautiful China" [37]. ZWC is a new concept to deepen the comprehensive management reform of solid waste at the city level, reduce and recycle solid waste to diminish its environmental impact, and minimize landfilling, and promote the construction of a "zero-waste society" [37]. ZWC can be viewed as an advanced urban development pattern that is guided by innovation, coordination, greenness, openness, and sharing to promote the formation of green development and lifestyle [37]. In 2019, the Ministry of Ecology and Environment (MEE) selected sixteen local authorities and districts to develop ZWC pilots [22]. The construction cycle of the ZWC spanned 2019 to 2020. In 2019, MEE released an indicator system to monitor the performance of ZWCs under construction. Focusing on solid waste reduction and resource utilization, this index system was designed to address five aspects: reduction of solid waste at the source, resource utilization, final disposal, guarantee capacity of institutional and market systems and people's sense of gain. In April 2022, MEE announced that 113 cities and autonomous prefectures would develop ZWC pilots in the 14th FYP period.

8 Discussion

8.1 Comparing CE Policies Across the FYPs

The CE goals changed during the three five-year plans. In the 11th FYP, the central government focused on the circular transformation of key industries and the construction of CE legislation. In the 12th FYP, the government engaged to provide fiscal, tax, and financial support, improve laws, regulations and standards, implementing a system of extended producer responsibility, formulating a catalogue of CE technologies and products, establishing a label system for recycled products, as well as establishing a systematic statistical evaluation system. In the 13th FYP, the central government attempted to establish a firmer link between production and household systems and accelerate the (re)utilization of waste resources. In the 14th FYP, circular production, green design, and clean production will be given extra attention. Moreover, resource utilization capacity would be enhanced.

Table A1 shows that over the periods of the 11th FYP and 12th FYP, the adopted economic policy instruments exceeded those of the legal policy instruments, although both were far more popular than other types of instruments. The total number of CE policies and CE policy instruments increased dramatically from 2006 to 2021. In the 13th FYP and 2021, all types of policy instruments were widely adopted. The trend of the 13th FYP continued in 2021, with an increase in the number of CE policies and all types of policy instruments. Legal and economic policy instruments still dominated policy mixes in the early years of the 14th FYP. In sum, interpreting the numerical figures in subtotal and the diversity of policy instruments issued and implemented, it is fair to state that a "policy accumulation" in the CE domain took hold over the 13th FYP and beyond into 2021 [20].

Except introducing a series of CE policies at the national level, China's government also proposed to have demonstration programs. These were proposed by the

State Council and mainly implemented by the NDRC and the MEE, which in turn selected specific cities or industries as demonstration pilots to implement CE. Table 2 shows the detailed descriptions and explanations for three CE demonstration programs from 2005 to 2022. The number of CE pilots increased since 12th FYP, especially after 2020 (regarding the ZWC pilots). We also found an accumulation process for the CE pilot projects. The names of CE pilots and ZWC pilots are listed in the Appendix (Table A2).

8.2 Drivers Behind China's CE Policy Accumulation

Three factors have driven the changes in China's CE policy adoption, innovation pilot schemes, and governance: (i) responses to economic growth and environmental degradation; (ii) institutions and differences in the division of functions between different central government ministries; and (iii) policy learning from foreign policy and practices. First, China's economic development had reached a stage where it started to pay attention to environmental protection. At the same time, the status of China's environmental deterioration had reached a point where the central government noticed that intervention was absolutely required. The need to transform production modes, the exploration of new pathways for economic growth, while pursuing high-quality development inspired the central government to adapt its strategy for future development patterns and adopt an explicit CE narrative. CE works as a carrier of "Ecological Civilization" to balance conflicts between environmental protection and economic growth. Secondly, competition across involved national government organizations influenced CE policy accumulation in China. With different ministries and committees actively involved in ecological civilization, many CE policies were issued and eventually implemented. For example, NDRC encouraged implementation of "CE pilots" and "National CE Pilot City (at the county level)". In contrast, MEE proposed two batches of ZWC pilots. This also reveals that CE policy accumulation has been advocated and driven by proactive state actors. First this occurred mostly at the national state level, but later they were also adopted by decentralized tiers of government, i.e. regions, cities, and counties. Thirdly, the central government adopted valuable lessons and practices on CE policy and management from other countries. For example, this happened when policy makers and knowledge institute researchers undertook site visits to the German Ruhr region to learn about industrial transformation and eco city construction.

9 Conclusion

This chapter has addressed the accumulation of CE policy in China from 2006–2022, by focusing on three dimensions: policy goals, policy instruments, and demonstration projects. By observing changes in the adoption of policy goals over this period it was

Table 2 CE demonstration projects by NDRC and MEE (see Appendix for full list of participating provinces, cities, districts and counties)

Title of CE pilot programs	CE pilots		National CE demonstration city (county)		ZWC pilots	
Ministry	NDRC		NDRC		MEE	
Year	2005	2007	2013	2015	2019	2022
Time scope	10th FYP	11th FYP	12th FYP		The end of 13th FYP	14th FYP
No. of pilots	10	17	40	62	16	113
Batches	I	II	I	II	I	ZWC in 14th FYP
Programs	CE pilots (provinces, cities, industries, fields, companies, industrial parks)		Cities (19); Counties (21)	Cities (26); Counties (36)	Cities (11); Districts or new town (3); Counties (2)	Cities (110); Autonomous prefecture (3)
	Provinces (3); Cities (7)	Provinces (4); Cities (13)				
Key elements (goals)	• Establish a system of laws, regulations and policies, a system of institutional and technological innovation, and an incentive and restraint mechanism for circular economy (2010) • Improve the efficiency of resource utilization, reduce the final amount of waste disposal, and build a large number of typical enterprises that meet the requirements of CE development • Promote green consumption and improve the system for recycling renewable resources • Build a batch of circular economy industrial (agricultural) parks, resource-conserving and environment-friendly cities		• Promote circular production modes • Promote green consumption models • Establish a system of resource recycling that covers the entire society • Improve the rate of resource production • Strengthen sustainable development capacities		• Establish an index system, comprehensive management and technical system for ZWCs • Improve the level of reduction and recycling of solid waste, including industrial solid waste, agricultural waste, household waste and hazardous waste • Foster enterprises for the recycling of solid waste • Promote ZWC demonstration models; promote the "zero-waste society"	

Sources [23, 28, 30, 32, 36, 37]

shown that the implementation of CE and environmental governance experienced a shift from focusing on production efficiency to decreasing consumption levels, to adopting a whole life cycle perspective. As shown in the 13th FYP and 14th FYP, policy goals did not just address the reduction of environmental destruction and pollutants' emission reduction, but also building a more livable and sustainable society. This evidences China's environmental protection turning into a more comprehensive, holistic mechanism. This study reveals that the CE policy accumulation in China is firmly tied to central government planning and strategy (ecological civilization), and indicates that the Chinese central government attached great importance to CE over 2006–2022.

Results show that the number of CE policy instruments increased dramatically, especially in the 13th FYP, a trend which (at least) continued until 2021. And with regard to the types of policy instruments adopted, a trend was observed moving away from economic policy instruments towards implementing instruments that work hierarchically and to exercise control. Increasing amounts of regulatory policy instruments and standardization, and in support some communicative and economic policy instruments were issued in the 13th FYP and the early 14th FYP. This shows that the CE policy is 'maturing', and that central government policy makers think that CE is only going to work if more planning, regulatory, and standardizing policies are used. However, at the same time an accumulation in CE and ZWC demonstration pilots (yet, organized in nation-wide programs) can be witnessed, indicating a combination of innovation-push and hierarchical policy instruments. This is clearly different from observations in Western world countries where economic policy instruments are typically combined with innovation policy to encourage transformative change.

Three factors indicated the accumulation of CE policy in China. First, China's economic development pushed the central government to start paying more attention to environmental protection, which led the central government to attaching greater importance to ecological and environmental protection, whilst combining resource conservation and environmental protection to establish the development goal of "Ecological Civilization". At this stage, the central government was also able to deal with environmental issues. To further proceed with the ecological civilization, a variety of ministries and commissions published increasing numbers of CE policies. Finally, policy learning and policy transfer from other countries also contributed to CE policy accumulation in China.

Over 2006–2022 different types of CE demonstration and pilot programs were issued across different FYPs. The results show that pilot projects encourage local governments to explore applicable solutions. Whereas the NDRC's pilot programs stressed improving resource efficiency and upgrading industrial technology, the MEE pilots emphasised environmental protection. This marked a shift from the solid waste oriented strategical approach by the Ministry of Environmental Protection to a whole life cycle oriented CE solutions approach. Whereas "CE Pilots" were implemented in Chinese provinces, cities, key industries, and companies, the "National CE pilot city" program mostly concerned CE policy implementation at the local level (i.e. in cities and counties), whilst the "Zero-Waste Cities" program emphasized specific waste management and treatment. This shows that the introduction and application of pilot programs play important roles in implementing a CE. Over 2006–2022, more pilot cities and regions were selected as demonstration projects or plots to demonstrate, experiment with and achieve CE and ZWC objectives. These developments contributed to further CE policy accumulation in China.

It suggests that future research pays more attention to how the implementation of CE policy takes place at the local level. Further research could also address comparative research designs into CE policy implementation in different regions because of considerable contextual differences among Chinese cities and regions, and in terms of CE adoption and governance. Moreover, the accumulation of CE policies and practices also can be researched in other countries, using in-depth case study research and later adopting comparative research designs.

Acknowledgments: This work was funded by the National Natural Science Foundation of China (grant number: 72404068). This work was also grateful for the support from Shenzhen City Overseas High-Level Talents Introduction Funding, Characteristic Innovation Projects of General Universities in Guangdong Province (2024WTSCX066).

Appendix

See Tables A1 and A2.

Table A1 Number of CE policy instruments over 2006–2021, classified as per policy instrument type [20]

FYP	Number of policies (subtotal)	No. of policy instruments				
		Legal	Economic	Network	Communicative	Total
11th	15	11	30	5	4	50
12th	22	31	47	31	17	126
13th	165	184	162	102	116	564
2021	83	199	115	40	78	432

Table A2 Three Chinese National circular economy city programs & related pilot initiatives

Circular Economy Pilots	2005	**Provinces (3)**: Liaoning; Jiangsu; Shandong
		Cities (7): Beijing; Shanghai; Chongqing; Ningbo; Tongling; Guiyang; Hebi
	2007	**Provinces (4)**: Shanxi; Zhejiang; Henan; Gansu
		Cities (13): Tianjin; Qingdao; Shenzhen; Handan; Fuxin; Baishan; Qitaihe; Huaibei; Pingxiang; Jingmen; Yulin; Shizuishan; Shihezi
National Circular Economy Pilot City (County)	2013	**Cities (19)**: Suzhou; Weifang; Guangzhou; Quzhou; Nanping; Chengde; Hebi; Loudi; Tongling; Huangshi; Jilin; Jincheng; Wuzhou; Jinchang; Wuhai; Guang'an; Puer; Dazu; Shangluo
		Counties (21): Ninghai; Yongkang; Xintai; Yanqing; Taixing; Shishi; Gaoyang; Jieshou; Guixi; Gucheng; Zixing; Diaobingshan; Boai; Xiaoyi; Tongwei Tiandong; Longli; Holingola; Geermu; Shanshan; Yimen
	2015	**Cities (22)**: Taizhou; Liaocheng; Xuzhou; Zhanjiang; Yangzhou; Jingmen; Luoyang; Shenyang; Xinxiang; Anshan; Ji'an; Fuyang; Changsha; Baotou; Liupanshui; Baiyin; Shizuishan; Lhasa; Tongren; Luzhou; Qujing; Liuzhou
		Districts (4): Chengyang; Jinghai; Qijiang; Hechuan
		Counties (36): Pingyuan; Anji; Zhaoyuan; Guangning; Luoding; Haining; Danyang; Anhua; Fengyang; Fengcheng; Zhijiang; Zhangshu; Fanchang; Qianjiang; Tonghe; Jianping; Taonan; Changge; Hancheng; Cengong; Linxia; Pujiang; Fuchuan Yao Autonomous County; 34 Regiment of the Second Division of Xinjiang Production and Construction Corps; Liangping; Xichong; Togtoh county; The 10th Regiment of the 1st Division of Xinjiang Production and Construction Corps; Manasi; Jingchuan; Yongning; Qingtongxia; Datong; Xiangyun
Zero-Waste Cities	2019	**Cities (11)**: Shenzhen; Baotou; Tongling; Weihai; Chongqing; Shaoxing; Sanya; Xuchang; Xuzhou; Panjin; Xining
		New Towns (3): Xiong'an New Area; Beijing Economic-Technological Development Area; Sino-Singapore Tianjin Eco-City

(continued)

Table A2 (continued)

Circular Economy Pilots	2005	**Provinces (3)**: Liaoning; Jiangsu; Shandong
		Cities (7): Beijing; Shanghai; Chongqing; Ningbo; Tongling; Guiyang; Hebi
	2007	**Provinces (4)**: Shanxi; Zhejiang; Henan; Gansu
		Cities (13): Tianjin; Qingdao; Shenzhen; Handan; Fuxin; Baishan; Qitaihe; Huaibei; Pingxiang; Jingmen; Yulin; Shizuishan; Shihezi
		Counties (2): Guangze; Ruijin
	2022	**Cities (110)** Beijing; Tianjin; Shanghai; Chongqing; Shijiazhuang; Tangshan; Baoding; Hengshui; Taiyuan; Jincheng; Hohhot; Baotou; Ordos; Shenyang; Dalian; Panjin; Changchun; Jilin; Harbin; Daqing; Yichun; Nanjing; Wuxi; Xuzhou; Changzhou; Suzhou; Huai'an; Zhenjiang; Taizhou; Suqian; Hangzhou; Ningbo; Wenzhou; Huzhou; Jiaxing; Shaoxing; Jinhua; Quzhou; Zhoushan; Taizhou; Lishui; Hefei; maanshan; Tongling; Fuzhou; Putian; Jiujiang; Ganzhou; Ji'an; Fuzhou; Jinan; Qingdao; Zibo; Dongying; Jining; Taian; Weihai; Liaocheng; Binzhou; Zhengzhou; Luoyang; Xuchang; Sanmenxia; Nanyang; Wuhan; Huangshi; Xiangyang; Yichang; Changsha; Zhangjiajie; Guangzhou; Shenzhen; Zhuhai; Fushan; Huizhou; Dongguan; Zhongshan; Jiangmen; Zhaoqing; Nanning; Liuzhou; Guilin; Haikou; Sanya; Chengdu; Zigong; Luzhou; Deyang; Mianyang; Yueshan; Xuanbin; Meishan; Guiyang; Anshun; Kunming; Yuxi; Puer; Lhasa; Shannan; Rikaze; Xi'an; Xianyang; Lanzhou; Jinchang; Tianshui; Xining; Yinchuan; Shizuishan; Urumqi; Karamay **Autonomous prefecture (3):** Dai Autonomous Prefecture of Xishuangbanna; Haixi Mongolian and Tibetan Autonomous Prefecture; Yushu Tibetan Autonomous Prefecture

Note NDRC = National Development and Reform Committee; MEE = Ministry of Ecology and Environment

References

1. Adam C, Hurka S, Knill C, Steinebach Y (2019) Policy accumulation and the democratic responsiveness trap. Cambridge University Press
2. CAoCE (2021) Develop circular economy and build zero-waste cty
3. Chien SS, Wu F (2011) Transformation of China's urban entrepreneurialism: case study of the city of Kunshan. Cross Current: East Asian History and Culture Review Inaugural Issue of Cross-Currents E-Journal (No. 1)
4. China Daily (2007, Oct 24) Ecological civilization. *The China Daily*
5. de Jong M, Yu C, Joss S, Wennersten R, Yu L, Zhang X, Ma X (2016) Eco city development in China: addressing the policy implementation challenge. J Clean Prod 134:31–41. https://doi.org/10.1016/j.jclepro.2016.03.083
6. European Commission (2020) A new circular economy action plan. European Commission. https://eur-lex.europa.eu/legal-content/EN/TXT/?uri=COM%3A2020%3A98%3AFIN

7. Fu Y, He C, Luo L (2021) Does the low-carbon city policy make a difference? Empirical evidence of the pilot scheme in China with DEA and PSM-DID. Ecol Ind 122:107238
8. Geng Y, Sarkis J, Ulgiati S, Zhang P (2013) Measuring China's circular economy. Science 339(6127):1526–1527
9. Goulder LH, Parry IW (2008) Instrument choice in environmental policy. Rev Environ Econ Pol 2(2):152–174
10. Halpern C (2010) Governing despite its instruments? Instrumentation in EU environmental policy. West Eur Polit 33(1):39–57. https://doi.org/10.1080/01402380903354064
11. Howlett M, Ramesh M, Perl A (2009) Studying public policy: policy cycles and policy subsystems. Oxford University Press. https://doi.org/10.1017/S0008423900007423
12. Huanqiu (2012) Promote ecological civilization progress. People's Daily. https://china.huanqiu.com/article/9CaKrnJy8E2
13. Justen A, Fearnley N, Givoni M, Macmillen J (2014) A process for designing policy packaging: ideals and realities. Transp Res Part A: Policy Pract 60:9–18
14. Kautto P, Lazarevic D (2020) Between a policy mix and a policy mess: Policy instruments and instrumentation for the circular economy. *Handbook of the Circular Economy*. Edward Elgar Publishing
15. Kern F, Howlett M (2009) Implementing transition management as policy reforms: a case study of the Dutch energy sector. Policy Sci 42:391–408
16. Knill C, Steinebach Y, Adam C, Hurka S (2020) Policy dismantling, accumulation and performance. *A Modern Guide to Public Policy*. Edward Elgar Publishing, pp 242–264. https://doi.org/10.4337/9781789904987.00025
17. Li L, Taeihagh A (2020) An in-depth analysis of the evolution of the policy mix for the sustainable energy transition in China from 1981 to 2020. Appl Energy 263:114611. https://doi.org/10.1016/j.apenergy.2020.114611
18. Ma W (2021) From city branding to urban transformation: How do Chinese cities implement city branding strategies? [Delft University of Technology, TPM]. https://doi.org/10.4233/uuid:c768cd19-f45e-4b1a-94e1-2a828d6cf175
19. Ma W, de Jong M, de Bruijne M, Mu R (2021) Mix and match: configuring different types of policy instruments to develop successful low carbon cities in China. J Clean Prod 282:125399. https://doi.org/10.1016/j.jclepro.2020.125399
20. Ma W, Hoppe T, de Jong M (2022) Policy accumulation in china: a longitudinal analysis of circular economy initiatives. Sust Prod Cons 34:490–504
21. McDowall W, Geng Y, Huang B, Barteková E, Bleischwitz R, Türkeli S, Kemp R, Doménech T (2017) Circular economy policies in China and Europe. J Ind Ecol 21(3):651–661. https://doi.org/10.1111/jiec.12597
22. MEE (2019a) Announcement on the List of Zero-Waste Pilot Cities. Ministry of Ecology and Environment of the People's Republic of China. http://www.mee.gov.cn/xxgk2018/xxgk/xxgk01/201905/t20190505_701858.html
23. MEE (2019b) Announcement on the release of the pilot list for the construction of 《zero-waste cities》. Ministry of Ecology and Environment of the People's Republic of China. http://www.mee.gov.cn/xxgk2018/xxgk/xxgk01/201905/t20190505_701858.html
24. Morseletto P (2020) Targets for a circular economy. Resour Conserv Recycl 153:104553. https://doi.org/10.1016/j.resconrec.2019.104553
25. Munnings C, Morgenstern RD, Wang Z, Liu X (2016) Assessing the design of three carbon trading pilot programs in China. Energy Policy 96:688–699
26. Nauwelaers C, Boekholk P, Mostert B, Cunningham P, Guy K, Hofer R, Rammer C (2009) Policy Mixes for R&D in Europe
27. NBoS (2022) Statistical communique of the national economic and social development of China in 2021. http://www.gov.cn/xinwen/2022-02/28/content_5676015.htm
28. NDRC (2005) Notice on the organization of the pilot work of circular economy (the first batch). National Development and Reform Commission. https://www.ndrc.gov.cn/fggz/hjyzy/fzxhjj/200510/t20051031_1203265.html

29. NDRC (2007) Notice on the issuance of circular economy evaluation index system. National Development and Reform Commission. https://www.ndrc.gov.cn/fggz/hjyzy/fzxhjj/200708/t20070814_1203269.html
30. NDRC (2013) National circular economy demonstration city (county) area in 2013. National Development and Reform Commission. https://www.ndrc.gov.cn/fggz/hjyzy/fzxhjj/201312/t20131231_1202951.html
31. NDRC (2015) List of national circular economy demonstration cities (counties) construction areas in 2015. National Development and Reform Commission. https://www.ndrc.gov.cn/fggz/hjyzy/fzxhjj/201512/t20151209_1203358.html
32. NDRC (2016) National circular economy demonstration city (county) area in 2016. National Development and Reform Commission. https://www.ndrc.gov.cn/fggz/hjyzy/fzxhjj/201601/t20160115_1203359.html
33. Reike D, Vermeulen WJ, Witjes S (2018) The circular economy: new or refurbished as CE 3.0?—exploring controversies in the conceptualization of the circular economy through a focus on history and resource value retention options. Resour Conserv Recycl 135:246–264. https://doi.org/10.1016/j.resconrec.2017.08.027
34. Ring I, Schröter-Schlaack C (2011) Instrument mixes for biodiversity policies. *Helmholtz Centre for Environmental Research*
35. Rogge KS, Reichardt K (2016) Policy mixes for sustainability transitions: an extended concept and framework for analysis. Res Policy 45(8):1620–1635
36. SC (2005) Several opinions of the state council on accelerating the development of circular economy. State Council. https://www.ndrc.gov.cn/fggz/hjyzy/fzxhjj/200510/t20051031_1203264.html
37. SC (2018) Notice on the Construction Plan of the 《Zero-Waste Pilot Cities》. State Council. http://www.gov.cn/zhengce/content/2019-01/21/content_5359620.htm
38. Tong X, Wang T, Chen Y, Wang Y (2018) Towards an inclusive circular economy: quantifying the spatial flows of e-waste through the informal sector in China. Resour Conserv Recycl 135:163–171
39. Tuominen A, Himanen V (2007) Assessing the interaction between transport policy targets and policy implementation—a finnish case study. Transp Policy 14(5):388–398
40. van Geet MT, Lenferink S, Leendertse W (2019) Policy design dynamics: fitting goals and instruments in transport infrastructure planning in the Netherlands. Policy Des Pract 2(4):324–358
41. Wang N, Lee JCK, Zhang J, Chen H, Li H (2018) Evaluation of Urban circular economy development: an empirical research of 40 cities in China. J Clean Prod 180:876–887. https://doi.org/10.1016/j.jclepro.2018.01.089
42. Wu F (2015) Planning for growth: urban and regional planning in China. Routledge
43. Xinhua (2017) Report to the 19th national congress of the communist party of China
44. Yeh AGO, Yang FF, Wang J (2015) Economic transition and urban transformation of China: the interplay of the state and the market. Urban Studies 52(15):2822–2848. https://doi.org/10.1177/0042098015597110
45. Zhu J, Fan C, Shi H, Shi L (2019) Efforts for a circular economy in China: a comprehensive review of policies. J Chinese MarkJeting, J Ind Ecol 23(1):110–118. https://doi.org/10.1111/jiec.12754
46. ZWIA (2018) Zero Waste. Zero Waste International Alliance. http://zwia.org/zero-waste-definition/

Open Access This chapter is licensed under the terms of the Creative Commons Attribution 4.0 International License (http://creativecommons.org/licenses/by/4.0/), which permits use, sharing, adaptation, distribution and reproduction in any medium or format, as long as you give appropriate credit to the original author(s) and the source, provide a link to the Creative Commons license and indicate if changes were made.

The images or other third party material in this chapter are included in the chapter's Creative Commons license, unless indicated otherwise in a credit line to the material. If material is not included in the chapter's Creative Commons license and your intended use is not permitted by statutory regulation or exceeds the permitted use, you will need to obtain permission directly from the copyright holder.

Building the Zero Waste City: A Half Century of Efforts in Beijing

Xin Tong

Abstract This chapter presents the development of the urban waste management system with the parallel transformation of the informal recycling sector in Beijing since marketization in the late 1970s. We illustrate the distressing challenge of waste as it is gradually exposed to urban governance in China. The municipality's efforts to remake the waste/recycling space into an "urban circular economic system" was depicted to highlight the context for the cities zero waste city initiatives. The implication points out the limits of problem definition in waste management, which focuses on the environment and resources, but excludes the migrant scavengers from the local citizenry regardless their efforts for inclusion by the city. This limitation of the definition of the problem leads to conflicting values on waste and recycling between various stakeholders who are involved when the system needs to be upgraded. The collaborations at the community level for making the new business model provide the opportunity for building an inclusive space for recycling activities in cities in China.

Keyword Urban waste management · Resources recycling · Beijing

1 Introduction

The informal recycling sector plays an important role in daily-life of cities in developing countries [11]. Unlike the urban waste management system in cities of the developed countries, which is highly dependent on capital intensive technical solutions with public expenditures, the informal recycling sector in developing countries provides an alternative approach of low-cost waste reduction through labor-intensive extraction of recyclable and reusable materials from mixed waste by scavengers and waste pickers. Despite the obvious health and environmental injustice issues related to these informal recycling activities [5], it has been declared to be an inevitable

X. Tong (✉)
College of Urban and Environmental Sciences, Peking University, Beijing, China
e-mail: tongxin@urban.pku.edu.cn

part of a feasible solution to the pressing challenge of urban solid waste management when accompanied with fast urbanization in developing countries [7]. The problem becomes how to integrate the informal sector into the modernization of the urban waste management system and how to benefit the people working in the recycling sector while at the same time adding the value of waste reduction to urban development [6].

The resurgence of informal recycling in urban China was one of the by-products of the market transition of the national economy during the 1980s, which was accompanied by the dramatic change in the economics of waste and the practices of recycling and reuse in people's daily lives [3]. As many studies have reported a positive role of urban scavenging and informal recycling in the transition scheme of waste in cities, this sector has received increasing attention in the policy arena since the late 1990s in China [1, 8]. Contrasting to developed countries where the waste issue has long been considered a crucial public affair in urban planning [2], waste management used to be quite marginal in Chinese cities. However, with the fast expansion of urbanization, conflicts between the demands for efficient waste disposal and the rise of NIMBY-signs in the public opinion increasingly challenged Chinese urban planners who are responsible for the location of waste disposal facilities within the cities. As the nexus of waste issues being revealed, it has been proven to be a distressing challenge to urban governance in contemporary China.

This chapter presents the case of Beijing, where the authorities have tried to integrate the informal recycling sector into its "urban circular economic system" since the early 2000s, with the inheritance of a multi-level collection network constructed during the centrally planned economy back in the 1950s. In parallel with the municipal efforts, Dongxiaokou, an unexpected "waste city" beyond the scope of official planning, grew explosively at the northern urban fringe. The complementary role of the informal system to the urban waste management turned into conflict when the municipal administration decided to demolish this "waste city".[1] Public debates aroused regarding the future development of urban waste management in China. The rise and fall of this "waste city" reflects the transition of the urban waste management system in China beyond the resources and environmental challenges.

2 The Changing Urban Waste Landscape

As is widely known, China used to have a comprehensive waste collection and recycling system under central-planning that was gradually constructed from the old junk world before 1949 to the essential part of the socialist economy at the end of the 1970s [3]. However, this system collapsed during the transition from a planned to a market economy beginning in the 1980s. A prevalent spatial phenomenon that

---

[1] Dongxiaokou received the name "waste city" (废城 feicheng) from the Chinese media beginning in 2014 mainly due to the debates regarding its removal during urban expansion. See, for example, http://news.china.com.cn/2014-06/19/content_32707970_2.htm.

occurred along with this structural change was the emergence of waste villages in the urban fringes of many cities in China. The activities in different waste villages varied from open dumping to garbage sorting of various recyclable goods. However, most of the waste villages shared common characteristics, such as the illegal occupation of rural land under the land management system in China and the concentration of migrant workers who made their living by scavenging, sorting, or recycling from urban discards. Such villages became places of tension and conflict in urban development because they provided a cheap solution for urban waste disposal but also created an ugly space that was host to poverty and misery near the city.

3   Building Formal Waste Management System

The recycling activity in waste villages was only one side of the coin in the transformation of waste conversion during the marketization in urban transition. For cities like Beijing, the waste system has been split into two worlds. One is formal waste disposal, which increasingly relies on public expenditure and capital intensive facilities, such as modern landfills or incineration. By constructing these modern waste disposal facilities, the municipality has successfully reduced open dumping areas, which used to surround the urban center. However, for recyclable garbage, it seems to be more difficult to keep the balance between private and public interests.

Before 2003, the Beijing Administration for Industry and Commerce was in charge of supervising and providing approvals to junk markets. Only entities that had received a certification could open recycling stations. Therefore, the certified recycling companies, which were mainly decedents from the central planning system, still had some privilege in the recycling system. Literally speaking, the urban scavengers could only sell their goods to the recycling stations opened by the certified recycling companies. The waste villages, which had concentrations of migrant scavengers, were close to those recycling stations, but markets for recycled products were gradually emerging within such villages, although they were by and large grey area practice.

4   The Transformation of Informal Recycling Sector

For example, Bajia was a village in Haidian District with an area of 1.6 square kilometers. In 1992, the former waste village in Erlizhuang, to the southeast of Bajia, was demolished due to urban expansion. At that time, there was a recycling station in Bajia run by the local government of Dongsheng Town. As a result, many urban scavengers who moved there rented county yards from local peasants and started junk sorting and recycling activities. In the middle of the 1990s, the number of rural migrants working in the recycling sector in Bajia was more than 5000, surpassing the number of local residents. Because more than 75% of these migrants came from

Fig. 1 The spatial flows of waste across regions

Henan province (a populous province in central China with large outflows of rural workers), scholars in the field of migrant labor studies named Bajia "Henan Village" [9]. Until now, the majority of junk buyers in Beijing are still migrants from Gushi, a small county in Henan.

However, in 2003, the requirement of administrative examination and approval for establishing a recycling market by individuals was cancelled[2] and was followed by an explosive growth of privately owned recycling markets around the city. There were more than 120 concentrated markets for recyclable goods in Beijing in 2004, with total land areas of up to 240 ha, among which only 24 markets had licenses from municipal authorities [4]. The majority of these sites were rented land from villages near the city, attracting large numbers of urban scavengers to conduct junk sorting nearby. Figure 1 shows the spatial flows of different waste after sorting according to the materials from waste villages in Beijing to the industrial users in other provinces. The waste villages in Beijing have played the role of hub of material flows between the consumer side to the production side.

---

[2] State Council [2002] No. 24, On cancellation of the first batch of administrative examination and approval.

5 The Production of Space in a "Waste City"

The first junk market was established in Dongxiaokou in 2003. Before that, there was a waste village in Wali, approximately 6 km to the south of Dongxiaokou, which was closer to the urban center. Wali was demolished to build the Olympic Park. A local peasant in Wali, who had been conducting junk sorting and recycling since the 1980s, decided to rent a 13 ha land plot from Dongxiaokou village and established a junk market. However, Dongxiaokou was not only a replicate of Wali. It had been transformed from a "waste village" into a "waste city".

The most obvious difference between a "waste village" and a "waste city" is the scale. After the opening of the first junk market, there arose 7 other markets founded by different bosses. Most of them came from Henan Province. They made money in Beijing from the scavenging and recycling business during the 1990s. From 2003 to 2008, the recycling sector was booming in China due to the increasing material demands for industrial growth. At the same time, a wave of urban (re)development in Beijing before the 2008 Olympics generated a good supply of secondary materials, especially metal scrap from dismantled buildings. Dongxiaokou won the largest share in the market upsurge relative to the other waste villages in Beijing. By 2009, it had absorbed one quarter of the recyclable goods generated by the city, with an annual value of transactions reaching 1 billion RMB (approximately 161 million USD). At its peak, it contained more than 1,000 family workshops, with over 30,000 people working and living in the area.

The economies of scale and scope were accompanied by organizational advantages. A flexible specialized network was emerging around the recycling markets. Materials were collected by the junk buyers around the city, sorted and traded in the markets, and then transported to recycling clusters in rural areas far away from urban centers for further processing into secondary materials or products. Each workshop in the market specialized in different types of junk recycling, such as metals, plastics, PET bottles, cotton fabrics, waste electronics, and furniture. As a local government officer said, "Here, you could find everything used in your home, being sorted and recycled by specialized workshops". The extensive linkages among the large number of family peddlers and workshops improved the efficiency of the whole recycling chain, especially for the recycling of complex products, such as waste electronics. The value of the discards varied greatly, and so did the proper method of recycling. It depended on careful evaluation, sorting, and transportation from the source of discard to the recycling sites. Through a disintegrated multi-level system, the discarded products were sorted according to their quality and value on the way to the recycling market from the dispersed households. Some of the products were refurbished and sold to poor migrants in the city or rural areas with lower incomes, and those that could not be reused were sorted and disassembled by different recyclers to be transported. The final goods were sorted according to their material composition and sold outside Beijing in bulk for further processing.

Self-organization was emerging to fit the evolving business structure. The migrant recyclers no longer rented houses or yards directly from local peasants. Without

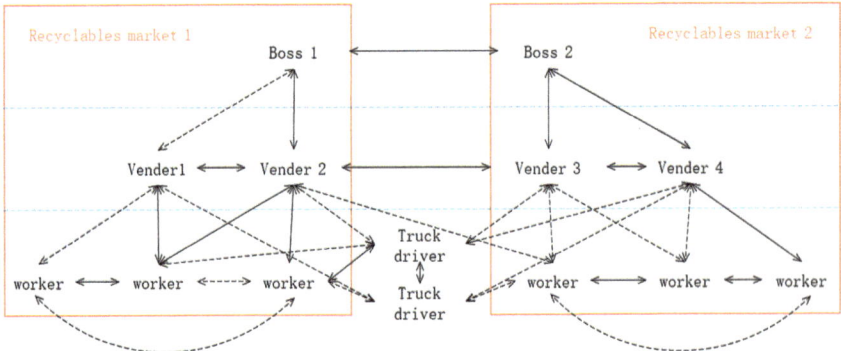

Fig. 2 The social network supporting the informal recycling sector

any official approval or planning from the government, the bosses of the recycling markets, who rented the land from the village, were trying to perform spatial planning autonomously. On the one hand, they knew the demands of the migrant junk buyers. After working in Beijing for years, many of the junk buyers had chosen to specialize in a segmented sector of recycling to establish business linkages with both up- and downstream customers and to accumulate capital for larger investments. Some of them even built their own recycling plants outside of Beijing. What they needed in Beijing was a small piece of land that they could live on with their family and in which they could sort and store their goods conveniently. Thus, the first market introduced the zoning strategy in spatial planning, dividing the space into several zones according to the category of recovered materials, and further split each zone into a number of small plots of 50 to 100 square meters. A small cottage to live in and a yard for junk sorting and storage were built on each plot. The market rented the plots to migrant recyclers, providing them a relatively comfortable place to stay. This arrangement was quite attractive to the migrant recyclers. Other markets that were established later followed this model. The social network based on the hometown identity supports the business linkages along the recycling chains (see Fig. 2).

On the other hand, the bosses of the markets knew the expectation of the government, that is, "Don't make any trouble for the city; no fires, no crimes…" Therefore, the markets provided some public services themselves. They established their own fire department, checked firetraps, and taught the recyclers in the market how to avoid fires. They invited a teacher from their hometown to establish an elementary school for the children of the recyclers. Because most people came from the same hometown in Henan, the markets also provided scheduled coach services between Beijing and Henan.

However, the concentration of urban recyclers in Dongxiaokou increasingly received negative exposure in the public media beginning in 2009.[3] Push-back

---

[3] Story of Waste Village. CCTV, 2014-01-14. http://v.ifeng.com/documentary/society/2014001/037f77e2-b690-4181-bf32-d46bf3d295f0.shtml.

came from local residents who complained about the dirty and congested environment. Some environmentalists, however, scaled up the local environmental concerns to address issues of uncontrolled flows of recyclable goods to other regions and inappropriate methods of processing hazardous materials.

Moreover, although junk market operators rented land from local villagers, the occupation of waste markets and workshops was generally illegal. According to Chinese Land Administration Law, rural land cannot be developed for activities other than agriculture without previous expropriation by the Municipal government and a change of its function to urban status.

Again, in 2011, Dongxiaokou was demolished. According to the official urban plan, this area would be developed into a large commercial and residential complex. Following the government decision to demolish this are, the majority of migrant recyclers accepted this decision and looked for new places to move to. However, the bosses, who have struggled to obtain considerable decent social status in the city by accumulating capital through recycling, tried to oppose the decision and raised their voices through different channels. One boss from Henan, who ran an iron and steel scraps market in Dongxiaokou, collaborated with a TV channel in Shanghai to produce a documentary on the waste village migrant worker's daily life, on their struggle to make a living, and on their contribution to the city and to resource recovery.[4] The video demonstrated his efforts to upgrade the recycling industry and improve the working conditions of his colleagues from the same hometown in Henan to Dongxiaokou in Beijing.

Furthermore, the efforts of the migrant recyclers received sympathy from Natural University, a local NGO working on environmental protection and community building in Beijing. They tried to raise public awareness of this marginal population, who provided low cost recycling services to the city but could hardly obtain a place of their own to live. Partly due to their efforts, some of the junk markets were temporarily preserved in 2013. The local government also promised to provide another place to accommodate the migrant recyclers. However, urban expansion finally won out over the efforts against demolition. The waste city in Dongxiaokou was removed almost completely before 2015. As most of the migrant recyclers left Dongxiaokou to find new places to stay, several other markets arose nearby, but more distant to the urban center. All of the markets are far smaller than Dongxiaokou. After the crash of the "waste city", "waste villages" returned.

---

[4] Real-25-hour. 2012. Billianaire Waste Village. RealMedia. http://v.163.com/jishi/V7MIUGGFJ/V8H2R1KN3.html Accessed August 2, 2015.

6 The Urban Circular Economic System for Zero-Waste City

Along with the rise and fall of the "waste city", the government had its own plan for the "Urban Circular Economic System". Since 2000, the municipal government has been trying to rebuild the hierarchical urban recycling system. It aims to reorganize the informal scavengers' network into an extensive set of community-based collection facilities covering all residential areas of the city to eliminate the crowded and untidy space of waste villages in urban fringes.

7 Tackling the Waste Problem with Inclusive Methods at the Community Level

In May 2000, 9 administration departments of the Beijing municipal government jointly promulgated an agenda concerning a pilot project on establishing community-based recycling systems (Jingzhenghui [2000] No. 8). Five of the eight inner city districts of Beijing, namely, Xicheng, Chaoyang, Haidian, Fengtai, and Xuanwu, were set as pilot regions. The actions included promoting the standardization of the logos, transportation vehicles, workers' uniforms, prices, and categories of recyclable goods, as well as measuring equipment in the recycling sector. Collection sites were deployed in all of the residential communities, each one serving 1000–1500 households [10].

The government also planned to build 10 formal recycling markets equipped with large automatic sorting and dismantling machines before 2003 to replace the dominant labor-intensive sorting activities that were prevalent in the waste villages. This scheme was supposed to cover 100% of recyclable products in the 5 pilot districts to regulate the uncontrolled flows of rags and to reduce the concentration of migrant workers in waste villages.

This plan can be considered to be an effort to rebuild a three-level recycling system similar to the one that existed under the central planning system. However, the emphasis went far beyond resource conservation and pursued a goal of producing a neat and clean image for the urban recycling sector. It provided detailed requirements on the community-based recycling facilities, including the design of cabins and logos, appearance of the vehicles, and so on. It also addressed the further processing of recyclable goods, including sorting, compressing, cleaning, and purification. A close-loop urban recycling system was designed to fit the image of a modern international metropolis.

The plan also tried to address social problems in the city, such as the unemployment of disadvantaged groups. It endowed local unemployed workers with priority for being hired as operators of community recycling facilities. The company could only hire qualified migrant laborers if they could not hire enough local residents. The plan created a set of rules and standards for the training, registration, social security, and human affairs management of the sector employees. It even suggested building a national qualification certificate system for the workers in this sector.

The certified recycling companies that descended from the former state-owned recycling stations carried out the plan. In 2008, 20 recycling companies of this type established more than 3000 community-based collection sites covering over 70% of the local resident population. They also built 13 sorting centers located in different districts. However, the operation of these collection sites still relied on migrant scavengers. Very few local residents wanted to enter this sector. Therefore, most of the collection sites were contracted to migrant scavengers, who had already built business contacts with the waste villages. Although the certified recycling companies provided cabins and uniforms to contracted collectors, they could hardly create a monopoly over such activities, either for them or their contracted collectors in the community. Additionally, the contracted collectors still had to compete with informal scavengers working door to door. Despite their contracts with certified recycling companies, the collectors would prefer to sell their collections to the market that could offer the highest price rather than sell them at a fixed price to the recycling company that they contracted with. The dynamics of secondary material markets gave the informal recyclers the expectations and opportunities to make more money from their hard work. Thus, the sorting centers of the formal recycling companies did not acquire much advantage over their rivals in the waste villages. Among them, the largest one, Dongxiaokou, was growing from a village into a city. There is a constant challenge to regulate the recycling market while keeping the economic vitality of the market competition.

8 Knowledge Gap on the Value of Recycling and Waste Reduction

The rise and fall of Dongxiaokou as a "waste city" shows the split mindset on waste between urban and rural, consumption and recycling, and public and private issues in contemporary China. Should we designate another place, or a set of places, to reproduce Dongxiaokou? Should we just let it go, so that waste villages can repeat themselves somewhere somehow? Or, can we find alternative and better solutions?

As increasing NIMBY-conflicts related to the construction of waste disposal facilities occurred in many cities in recent years, the distressing challenge of waste was gradually exposed to urban governance in China. However, the problem is generally seen as a tension between state authority and the growing civil society within the middle class in cities. The informal recycling sectors are, by and large, not included as empowered stakeholders in the debate; however, they are seen as either a symptom of the problem or a tool that could be an option for solving the problem. The story of Dongxiaokou has proven that the definition of the problem was indeed the cause of the problem. We need to reformulate the governance structure in the adaptation strategy to tackle it.

The coping strategy at the municipal level has shifted from authoritative to competitive, which means that bottom-up rationalization was permitted for the different stakeholders at community level to test better solutions. The municipal authority, the urban residents, and the informal recyclers obviously have conflicting values regarding recycling and waste reduction. The urban residents expect a convenient and affordable waste disposal system, and it needs to adhere to "Not In My Back Yard". The government expects a modern and clean waste management system that does not rely on a large number of migrant laborers so that ugly waste villages may eventually disappear. The informal recyclers just want to make a living in the junk business in their cities.

Without including the informal recycling sector as a stakeholder in the pursuit of a solution, it seems that an easy way for the government to copy the model of developed countries is to build more capital intensive disposal facilities with public expenditure. The ever increasing public funds and subsidies to the formal waste management system actually concentrated the power of decision making into the hands of a few people at the municipal level. Technical solutions, such as incineration and landfills, granted authority to a small number of professional experts. However, competing points of view cannot be eliminated, neither from the NIMBY-protests nor from the recycling advocates.

Even without a great deal of influence in the public decision making process, the informal recycling sector as a competitive solution to waste reduction was largely successful when the revenue from the resource recovery exceeded the labor cost. This success was achieved by an extensive network of disaggregated actors along the recycling chain. As in the case of Dongxiaokou, the migrant recyclers were willing to invest in upgrades based on their knowledge of the business. One of our interviewees recalled his struggles in the city. Despite all of his miserable experiences, he was very proud of his current business. He owed his success to one professor who researched the recycling of waste plastics many years ago. He was only a street picker in the city when he met the professor during his field work. From that professor, he learned about the business opportunities that were available in plastic recycling. At the time of our interview, he had a small spot in the recycling market in Dongxiaokou and a plastic recycling plant outside of Beijing. He was quite satisfied with the living conditions in Dongxiaokou. He took it for granted that if the government designated 4 pieces of land to build concentrated recycling markets, such as Dongxiaokou, on the urban fringe in each direction of the city, the migrant recyclers would pour in.

However, the owners of small workshop could only depend on somebody else to build the market for them. They didn't have the capability to directly invest in building new markets. A boss of the recycling market in Dongxiaokou had the investment capability. He closed his market in Dongxiaokou in 2013, and successfully rented a piece of land from a village nearby. He designed a new market according to his understanding on how to upgrade the facilities to satisfy both the demands of migrant recyclers, and the expectation of the government. He had a son who had grown up in Dongxiaokou and now worked in a real estate company in Beijing after graduating from college. Both father and son were trying to lobby the local government to permit them to open their new recycling market. The son said that if the new market received approval from the government, he would like to withdraw from the real estate business and devote his time to the recycling business. However, according to the official land use plan, the piece of land they rented could not be used to build a recycling facility.

The challenge of initiating collaborative action towards tackling the problem is sharing knowledge among stakeholders, especially those with competing values. Indeed, the competition between the formal and informal sectors created a confrontational environment that discouraged knowledge sharing and collaboration, as the former threatened the profitability of the investment by the latter. For example, the certified recycling companies were interested in upgrading their recycling facilities as well. Their new system design was closer to the expectation of the government. They planned for a large investment on automatic sorting and processing equipment, large storehouses, reverse vending machines for bottles, and reverse logistic systems for collection from households. However, to guarantee investment returns, they asked for an exclusive franchise or subsidy from the government.

The conflicts of interest among stakeholders came to a dead lock at the municipal level. The value of waste reduction has been appreciated at all levels; however, without grounded tools for accountability, the informal recycling sector can hardly benefit from their contribution on waste reduction. This problem intensified when the market value of recovered materials decreased, as economic growth has slowed in China in recent years.

9 Community as the Common Place for Action

The changing mind-set on waste has shifted the focus of action to the local level (Aspinwall and Cain 1997). Following this trend is the rise of community-oriented tools in recycling as well as the promotion of other sustainable behavior. By bringing society back in, the grassroots initiatives that are being implemented at the community level could provide a better platform for collaborative capacity that is built from the bottom up.

By contracting migrant scavengers in residential communities to work in recycling cabins, the top-down deployed resource recovery facilities of the "urban circular economic system" in Beijing have actually benefited from the growing revenue from recovered materials through their extensive linkages with the informal recycling sector during the upsurge of the recycling business from 2003 to 2007, in which Dongxiaokou was the tip of iceberg. However, despite the thriving linkages among the informal recycling network, the linkage between the recycling business and the households in residential communities decreased. The spatial segmentation between the waste village and the city is the result of social segmentation between bottom rural migrants and the middle class in the city. The solution may not lay in the waste village itself, but in community changes throughout the city. It is not enough to include the informal recyclers as an attachment to the recycling facilities deployed in the community, but they need to be active enablers in sustainable community initiatives.

The local governments of cities in China has been aware of the necessity to improve the garbage sorting by households at the source of generation. Innovative solutions at community level have been practiced in many cities in recent years, for example, the Ala recycling credits in Shanghai[5] or the community recycling initiative by Green Earth in Chengdu.[6] However, these programs generally depend on subsidies from the government. The introduction of the principle of Extended Producer Responsibility in waste management in China provides an alternative financial mechanism for certain categories of waste flows other than local public expenditure. No matter where the financial support comes from, it is crucial to track the volume and quality of the sorting process as a reliable measurement on waste reduction and resource recovery. It will be even better if the system could be flexible enough to fit the business routine of the informal sector. We joined this wave of experimental efforts by initiating a recycling program in Hongfuyuan, a residential community in Changping District of Beijing, not far away from Dongxiaokou. We use the technical solution of Green Earth, providing incentives to the participant households according to their sorting performance. And we collaborated with a family of informal junk-buyers in the community during the experiment. Since December 2013, the scale of the experiment has only reached around 500 households, roughly 5% of the whole community. However, it tests both the practicability of technical solution as well as the possibility to include the informal sector at the community level.

10 Conclusions: Building an Inclusive Space for a Circular Economy

This chapter presented our observations on the bottom-up rationalization of informal recycling in Dongxiaokou, whose dramatic growth led to the development of a 'waste city' from a large number of 'waste villages', in tandem with the top-down efforts of the municipal government to build an urban circular economic system. The crash of Dongxiaokou shows the big gap between expectations of the government on a clean and tidy recycling space and the demands of migrant recyclers struggling to survive in the city they are working for. The difficulty to keep a vital recycling sector while achieving higher environmental standards requires a reframing of the issue of social inclusion of human beings in the material-oriented recycling businesses.

---

[5] AlaHB, a recyclable transaction platform supported by the local state-owned recycling company, initiated their community-based recycling program for e-waste since 2010. See details on http://www.alahb.com/.

[6] Green Earth, a community-based garbage sorting company in Chengdu, established an information system for the residential community providing incentives for registered members according to their performance in garbage sorting by tracking their discarding behavior with code bar on garbage bag. See details on http://www.lvsediqiu.com/.

First, the problem definition in waste discourse should go beyond the physical deployment of recycling sites or waste disposal facilities. Dongxiaokou was neither like the modern recycling sites modelled from developed countries nor the same as the waste villages that had already disappeared. It provided a method of upgrading the recycling section by correctly responding to the structural change of the informal recycling sector and found a balance between the demands of migrant recyclers and the expectations of the government in the early 2000s. After the demolition of Dongxiaokou, the gap between the demands of informal recyclers and the expectations of the government widened. The city can hardly replicate another Dongxiaokou just by designating a place for it.

Second, conflicting values on what the new space should be like reflect the deep urban–rural segmentation in China. The authoritative strategy for capital intensive waste disposal facilities met increasing resistance from both NIMBY-protests and recycling advocates. Using the informal recycling sector as a competitive strategy for the formal waste management system is also losing its economic feasibility. To form a collaborative strategy among the state authority, the middle class urban residents, and the informal recyclers, the migrant workers with local citizenship need to be empowered before a shared understanding can be reached among the different stakeholders.

Third, the community-based recycling system built under the "urban circular economic system" in Beijing was the correct effort towards implementing better solutions. It includes some of the migrant scavengers in the local common space of the neighborhood. During the upsurge of the recycling business from 2003 to 2007, this system actually benefited from the growing revenue from recovered materials through its intensive linkages with the whole informal recycling sector, including the dramatically growing "waste city". However, because the market for recyclables shrinks as the whole economic growth slows down, the whole system needs support from broad collaborative actions. The action should be grounded in households, neighborhoods, and communities so that everybody can take part and where everybody can see the result of their collective actions. In 2018, Aifenlei, as one of the cases of a door-to-door collection model addressed in this chapter, was extablished in Changping district, where Dongxiaokou was located. The founders of this company are father and son. The father used to do recycling business in Dongxiaokou, and closed the business after the demolition. With inspiration for a new business model, the son quit his job at a real estate company, and started the recycling business with his father in Changping again. They are now representative as the enablers for the building of an inclusive space for recycling in Communities in Beijing.

References

1. Chi X, Streicher-Porte M, Wang MYL, Reuter MA (2011) Informal electronic waste recycling: A sector review with special focus on China. Waste Manage 31:731–742
2. Davoudi S (2000) Planning for waste management: changing discourses and institutional relationships. Progress in Planning 53:165–216
3. Goldstein J (2006) The remains of the everyday: one hundred years of recycling in Beijing. In: Dong MY, Goldstein J (Eds) *Everyday modernity in China*. University of Washington Press, Seattle
4. Liu Y, Li X, Xu Y (2008) The planning of recycling system in Beijing. In: Environment and sustainable development (In Chinese) (pp 1–2)
5. Medina M (1997) Informal recycling and collection of solid wastes in developing countries: issues and opportunities. UNU/IAS working paper. UNU/IAS
6. Scheinberg A, Spies S, Simpson MH, Mol AP (2011) Assessing urban recycling in low-and middle-income countries: Building on modernised mixtures. Habitat Int 35:188–198
7. Sembiring E, Nitivattananon V (2010) Sustainable solid waste management toward an inclusive society: Integration of the informal sector. Resour Conserv Recycl 54:802–809
8. Steuer B, Ramusch R, Part F, Salhofer S (2017) Analysis of the value chain and network structure of informal waste recycling in Beijing, China. Resour Conserv Recycl 117:137–150
9. Tang C, Feng X (2000) Stratification of rural migrant labor in Henan Village. Sociological studies (in Chinese) (pp 72–85)
10. Wang J, Han L, Li S (2008) The collection system for residential recyclables in communities in Haidian District, Beijing: A possible approach for China recycling. Waste Manage 28:1672–1680
11. Wilson DC, Velis C, Cheeseman C (2006) Role of informal sector recycling in waste management in developing countries. Habitat Int 30:797–808

Open Access This chapter is licensed under the terms of the Creative Commons Attribution 4.0 International License (http://creativecommons.org/licenses/by/4.0/), which permits use, sharing, adaptation, distribution and reproduction in any medium or format, as long as you give appropriate credit to the original author(s) and the source, provide a link to the Creative Commons license and indicate if changes were made.

The images or other third party material in this chapter are included in the chapter's Creative Commons license, unless indicated otherwise in a credit line to the material. If material is not included in the chapter's Creative Commons license and your intended use is not permitted by statutory regulation or exceeds the permitted use, you will need to obtain permission directly from the copyright holder.

Household Renovation Waste in the Netherlands: Mapping the Social Side of Waste Flows

Daan Schraven, Kai Vaessen, Zhaowen Liu, and Tong Wang

Abstract The transition to a circular economy necessitates cities to effectively manage their waste flows by capturing and redirecting them within the municipal domain. Traditional approaches to controlling waste flows have primarily relied on quantitative methods, such as material flow analyses. These methods excel in mapping the quantitative aspects of the materials and visualizing their sequential movements. However, recent advancements have highlighted the significant role of human behavior in shaping waste flows. For example, the way that citizens sort their waste determines the components and calorific value of the flows and directly impacts the circular rerouting challenges within the city. This paper argues that enhancing waste management practices in a circular economy requires analytical tools capable of incorporating the social dimension. Based on this premise, a novel approach termed the Waste Journey is developed and tested using a case study on household renovation waste in the Netherlands. Various methodological options to map the case are discussed and an initial framework for the Waste Journey is then proposed. The study emphasizes the influence of social processes on waste handling and offers a comprehensive means in which these processes can be mapped to effectively address challenges toward zero-waste cities.

Keywords Waste management · Circular economy · Social engagement · Waste journey

D. Schraven (✉) · T. Wang
Faculty of Architecture and the Built Environment, Delft University of Technology, Delft, The Netherlands
e-mail: D.F.J.Schraven@tudelft.nl

K. Vaessen · Z. Liu
Faculty of Civil Engineering and Geosciences, Delft University of Technology, Delft, The Netherlands

1 Introduction

Besides greenhouse gas emissions, it is forecasted that cities will contribute to 50% of global waste generation by 2050 [4]. Next to the forecasted urbanization of 68% of the world population living in an urban area by 2050, the infrastructure that needs to handle and cope with the city waste collection will be overburdened and inadequate [60]. In response, among others, to this global societal challenge, the United Nations formulated the Sustainable Development Goals (SDGs) in 2015, as a global call to action. Within this grander scheme, the Dutch government formulated their strategic ambitions to a full-scale transition to a Circular Economy (CE) by 2050 [34]. In this CE strategy the Dutch government gives waste management a crucial role, through the publication of the National Waste Management Plan 3, or NWMP3 in short. In the NWMP3 the Dutch government seeks to (1) limit the generation of waste, (2) restrict the burden of production chains on the environment, and (3) optimize the use of waste materials [44].

To date the academic literature has been addressing a wide variety of problem domains on waste management in a quantitative fashion. For example, Wang et al. [65] studied the flow of waste from decorations and renovations from the generation of waste to the end-state (e.g., recycling or land-filling) and estimated the different waste generation rates for different materials. This quantitative approach can also be seen in the study of Ding et al. [8], in which the waste generation rates of renovation waste are determined based on the reasons for renovating a house (e.g., carpentry or painting). Other studies focus on the role of informal waste collectors in a material flow analysis of the television waste management process [56]. Other topics seen in recent studies are waste generation rates, waste collection methods, waste management, sustainable development goals, and waste composition studies [22]. As these studies primarily focus on quantitatively addressing how the waste follows a specific stream from generation to its end-state, they generally lack attention on the social side. As a result, they miss out on the role and impact of different actors as they interact with waste, which has become the key of realizing the objectives proposed by the NWMP3. In this plan the Dutch government emphasizes waste prevention and reuse, and it requires the broad participation of actors, such as households and businesses, to change their production and consumption behaviors at early stages. Thus, the management of the waste stream is no longer only about the quantity, but also the "quality" of the waste stream (i.e. how well it is separated for further treatment). Furthermore, waste treatment has been recognized not only as an environmental challenge, but a complex economic-social and systemic issue [28]. Therefore, an actual focus on the inclusion of actors in the waste management process is needed to unlock the opportunity for shifting perceptions of the overall performance of the waste management system [68], most notably on the aspects of circularity and inclusiveness.

This chapter aims to address this issue by answering the question: *"How to identify barriers and opportunities for realizing inclusiveness and circularity in a waste management process?"* As Chapter 2, 3 and 4 of this book elaborated, in the context of waste management, we define inclusiveness as "everyone in the society can fully participate in and benefit from waste-related activities equally", and circularity as "using R-strategies in the socio-economic system to replace the linear economy". In the following sections, the state of the art in methodologies for waste management, circularity, and inclusion are reviewed in Sect. 2, and an approach—Waste Journeys—is proposed. After the methodology is developed in Sect. 3 it is applied to a Dutch case of household renovation waste (HRW) in Sect. 4. Section 5 discusses the implications of this method and Sect. 6 offers generic conclusions.

2 Theoretical Underpinnings

In this section of the paper, we develop the theoretical framing regarding the needs of capturing data on the social side of a waste management process. First, we review the generic background to CE. Then we review the generic inclusiveness background and inclusive waste management. Thirdly, we discuss the generic traits of material flow analysis (MFA), their shortcomings in light of inclusiveness and how new traits from the customer journey and other theoretical underpinnings can offer a new way to make waste flow analysis more inclusive by focusing on the social side. Finally, we finish with a conceptualization of this method, which we coined the Waste Journey.

2.1 General: Circular Economy Background

Some materials that are commonly used are depleting. In fact, the extraction of these materials should be limited to zero [23]. Until recent days, most economies in the world are still operating based on linear principles in which raw materials are used to produce products, which are then disposed of as waste at the end of their life. This linear economy is also known as the "take, make, and dispose" economy and has many negative effects on different aspects (e.g. environment) that, in the end, can threaten the survival of humanity and needs to change fundamentally [17]. A consensus is reached that a more sustainable economic model is needed.

Within this overarching debate, the CE concept gained much momentum in the academic literature in recent years [27]. The concept itself originates in different scientific disciplines, such as environmental economics and industrial ecology [33]. In environmental economics, Pearce and Turner [38] described that a transition would at some point take place from an old open-ended economic system to a CE system, following the laws of thermodynamics. In this field, some economists described three economic functions of nature (providing resources, life support system, and placing waste and emissions), which would come at a price and market, and therefore needed reflection in economic terms like the price of a product [5, 38]. Notably, pricing nature's products is an incomplete approach that does not account for the intricate, interdependent dynamics among the single subsystems of the larger human–environment ecosystem. In industrial ecology, the CE system stems from the examinations of the industrial system and its environment, making up a coupled ecosystem which can be described by flows of material, energy, and information [17]. Against these backgrounds, the CE is positioned as a holistic view of processes and systems.

In practice, many regions have recognized circular economy transition as one of the key strategies for sustainable development. In 2008, the Chinese government released the law of "Promotion of Circular Economy", defining CE as the practice of reducing, reusing, and recycling in production, distribution, and consumption [30]. The European Union followed in 2015, in their action plan describing CE as "an economy where the value of products, materials, and resources is maintained in the economy for as long as possible, and the generation of waste minimized" [11, p. 2]. In March 2020, the European Commission embraced a new Circular Economy Action Plan (CEAP), a pivotal component of the European Green Deal, which outlines Europe's blueprint for sustainable growth [12]. This shift towards a circular economy within the EU not only alleviates the strain on natural resources but also fosters sustainable economic expansion and job creation. Moreover, it stands as a fundamental requirement for attaining the EU's 2050 climate neutrality objective and reversing the trend of biodiversity decline. This new action plan introduces a series of initiatives spanning the entire life cycle of products. To realize these goals, the new action plan incorporates a blend of legislative and non-legislative measures, pinpointing areas where EU-level intervention offers substantial added value.

Kirchherr et al. [27] analyzed 95 definitions and found that the core principles of the circular economy include the R-framework, waste hierarchy, and systems thinking. The 10R framework developed by Potting et al. [40] presents ten strategies (refuse, rethink, reduce, reuse, repair, refurbish, remanufacture, repurpose, recycle, recover) in a hierarchical order that contribute to the CE. In the European waste directive, the Waste Hierarchy set a top-down prioritization of waste treatment strategies: prevention, preparing for reuse, recycling, recovery, and disposal [10]. The final core principle is systems thinking, which underscores the necessity of a comprehensive system overhaul to transition to a circular economy. A circular economy involves a complex system in which individuals, businesses, and ecosystems are interconnected [28].

2.2 General Inclusiveness Background and Inclusive Waste Management

Inclusiveness is a recent term that is derived from inclusion. As the concept of inclusion is difficult to define, one could also look at what is not inclusion. Its antonym, exclusion, could be defined as 'a state in which individuals cannot fully participate in their political, economic, and social lives' [59]. A former minister from the government of France spoke about exclusion as "mentally and physically handicapped, suicidal people, aged invalids, abused children, substance abusers, delinquents, single parents, multi-problem households, marginal, asocial persons, and other social 'misfits'" [42, p. 162). This term emerged after the European social welfare crisis and formed the basis for social studies on exclusion.

The UN coined the term inclusive society as a "society for all" (United Nations, n.d.) and described it as a society providing mechanisms for people so that they can actively participate (in all political, economic, and social dimensions) and are assured that they receive the same opportunities, regardless of their background (United Nations, n.d.). According to a study by Gerometta et al. [16] inclusion in social relations is formed by interdependence (in the form of division of labor and social networks) and participation (in the form of ability to consume, political power and education).

In the city context, inclusiveness is adopted in the thought of the inclusive city. The World Bank conceptualizes that an inclusive city exists "in a complex web of multiple spatial, social and economic factors." Chapter 3 of this book gave six interwoven dimensions of inclusion in the city context, namely the:

- spatial dimension, which ensures equal access to housing, services, and infrastructure;
- social dimension, which focuses on equal access to social resources and creates ownership of these social resources;
- economic dimension, which allows everybody, especially the disadvantaged, to share in rising prosperity, including labor and welfare services;
- environmental dimension, which addresses the current generation's environment and natural resource demands without sacrificing future generations' interests;
- political dimension, which grants citizens equal political rights and obligations and ensures a non-discriminatory relationship between the state and citizens;
- cultural dimension, which means cultural heterogeneity and diversity of cities should be taken into account in the urban policy-making process, so that personal cultural belonging can be respected in a society where ethics and common values are guidelines.

To summarize, it can be conceptualized that inclusiveness is the degree to which all people, including the most disadvantaged, can actively participate in, make use of, benefit from, and are affected by various aspects of their lives within spatial, political, economic, environmental, social, and cultural dimensions. In this way inclusiveness is also affecting the waste management as this is an important service in cities. It requires participation and cooperation between the different actors, which includes the households, recycling companies, and waste collectors, among others. If any of these actors does not experience the same level of inclusiveness as others, it can disrupt the balance and functionality of the system.

More specifically, in less developed countries, inclusiveness in waste management is regarded as the incorporation of traditional informal waste collectors in the system [7, 12, 19–21, 32, 37, 46, 49, 51, 54]. Most of these studies observe that many cities around the world are modernizing their waste management system, and that the technocratic approach actually undermines the role of informal waste pickers and collectors [69], since these actors are neither included nor intended to participate in these modernized processes for various reasons.

To alleviate this, the design of a waste management system needs to include these actors in the decision-making process and ensure that the design can create mutual benefits for all these actors. In doing so, qualitative indicators were constructed for user and for provider inclusiveness [57]. User inclusiveness can be defined as: "the extent to which users have a say in the waste management services in a city" [3, p. 257]. Provider inclusiveness is "the degree to which service providers are involved in waste management planning and implementation processes" (ibidem, p. 257).

Making the waste management process and system inclusive ensures that various actors can play crucial roles in effective waste management by cities, supporting related circular economy ambitions [24]. In that sense, informal stakeholders should be seen as part of an inclusive circular system, contributing to long-term urban sustainability [62].

2.3 The Analytical Traits of Methodologies for Inclusive Waste Flow Analysis

One of the methods that is used most often in waste management studies is the Material Flow Analysis [14, 68]. MFA is an accounting approach to substance flows entering, remaining within, and leaving a defined socio-economic system. The quantification focuses on the waste flows and their trajectory through various parts of the waste management process. Such a quantification is particularly useful for assessing the level of circularity of a flow/system, by quantifying which flows remain within a system vis-a-vis which leave it, indicating the extent of circularity of that very system [14, 68]. One notable shortcoming of MFA is its limited emphasis on the role and the interaction of actors in the waste management process. This neglect can result in an incomplete understanding of the complexities and challenges associated with waste management practices. For instance, MFA may not adequately capture the role of informal waste pickers who are often marginalized and operating outside formal waste management systems, but still contribute to the recycling process [52]. In addition, MFA misses qualitative features of waste systems that make the MFA happen, i.e., routines and exchanges between stakeholders as well as their value ideals and norms that motivate them to engage in any consumption, recovery, or disposal pattern.

Next to MFA, there is a method called the Wasteaware benchmark. It can be used to identify weak spots in a solid waste management system of a city [3], as well as to measure the performance of a waste system allowing for a comparison between cities [66]. The data that feed the indicators of the benchmark are intended to be easily usable. However, conclusions regarding the results are limited to merely comparing the quantitative features between processes. It lacks a proper qualitative interpretation and the role of actors (users and general public) have with the performance of the waste management system. Still, the Wasteaware benchmarks can be used as criteria for the degree of user and provider inclusiveness.

Clearly both MFA and Wasteaware benchmark have some limitations when it comes to the social side of a waste flow analysis. A recently applied method managed to take on board inclusiveness and circularity in waste management through the theory of planned behavior [39]. This is a framework explaining how intentions and behavior play a crucial role in a person's decision-making process. The behavior of an individual is formed through different constructs like attitude, subjective norm, and perceived behavioral control [2]. Pongpunpurt et al. [39] extended the theory by adding a situational and knowledge factor and therewith used it to capture people's intentions for household separation at the source. With this, the researchers set up hypotheses about the factors to explain the intention of waste separation by an actor, and found that residents in Bang Chalong, Thailand, were positively and most significantly influenced in their waste separation by their knowledge and subjective norms.

Figure 1 shows the schematic representation of the theory of planned behavior and its behavioral factors. In this theory, attitude is used for arguments that relate to the attitude the actor has towards separating or recycling waste. The subjective norm is used for arguments that relate to any form of social pressure, like separation is desired by the community. Perceived behavioral control refers to the reflection of an actor on their own ability and willingness to perform a given behavior, like easy or hard to separate waste. The situational factors relate to the elements that form the actor's contextual situation, for example, the location or facilities that allow or compel to separate waste. Finally, knowledge refers to the actor's knowledge of sorting waste, like knowing how to separate paper from a bottle.

Interestingly, this study by Pongpunpurt et al. [39] set up an attractive way to include the social side of the waste management process, however, it only went as far as targeting one node, whilst the waste management process consists of multiple nodes through which waste flows. This insight therefore gives a few implications for a theoretical development for inclusive waste management. First, it is observed here

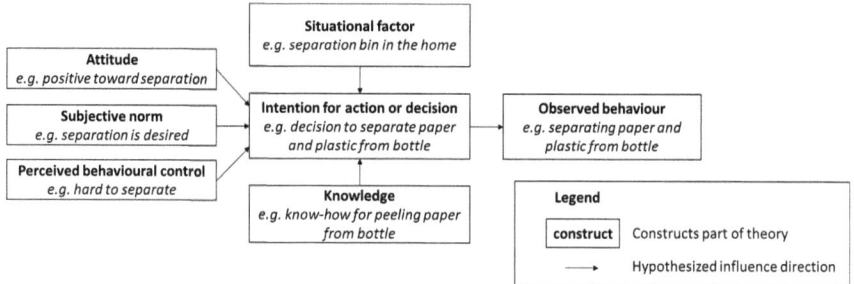

Fig. 1 Schematic representation of theory of planned behavior with examples inspired by Pongpunpurt et al. [39]

that this theory could serve as a basis to extract the social engagement of one actor in a chain of multiple nodes. Second, it shows the potential of the theory of planned behavior to describe the nodes along a waste flow as the parts where actors' decisions and behavior will influence the path of the waste through the system. Thirdly, this level of analysis could unlock the potential barriers and opportunities that these decisions cause in the system to realize inclusiveness and circularity.

Following from the aforementioned implications, it is clearly shown that experiences of an actor have a strong impetus on how decisions are made, but usually at just one place in the flow of the waste. However, in order to improve the waste management at a system level, an effective analysis would have to rely on methods that can (1) view the underlying behavioral factors of the theory of planned behavior and also (2) analyze this across the entire path of multiple places in the flow that waste is travelling, (3) and therefore be flexible enough to follow the direction of the decisions by each of these individual actors. In this sense, measures for improving the circularity would have to be understood during the entire flow, whilst the screws to turn would have to be understood through the inclusive actor level. Therefore, a method which is able to follow the actor involvement from the point of view of the waste would have to be facilitated to analyze actor-level measures for system level problems.

This extraordinary set of criteria challenged the researchers to think beyond the usual methodology repertoire in waste management studies. As a matter of mere unconventional interest, it was taken as a challenge to look for a method that would fit well inside these set criteria. Following from that it is argued that the design sciences offered an unconventional approach, called the Customer Journey, that suits the needs for the requirements as described above.

The customer journey knows various definitions. It is referred to as a process that a customer goes through, where the stages and touchpoints collectively make up the customer experience [55]. Also, Terra and Casais [53] define it as a sequence of events in which the customer goes to look for information and interacts with a product or a service. It looks at the complete sum of the customers' experience on interactions with a brand or product [41] including the pre-purchase, purchase, and post-purchase stages [14].

A few analytical traits are noteworthy here. The focus is on the sequence of events, also called points of interactions, from mostly the customers' perspective [13, 50]. At these interaction points, the analysis looks at how the customer acts and makes decisions regarding the use of the product or service and treats these as separate event stages [48]. In this way the journey that a customer goes through can be evaluated retrospectively [47] and can be used to understand where the product or service creates value for the customer [61]. This way the customer journey helps to identify barriers and opportunities that can help to improve the process of the customers experience along the journey. It should be noted that the customer journey, in some instances, can also be focused on the perspective of the product or service provider to identify certain issues.

Although the customer journey is not used in waste management studies, it proves to be useful for evaluating the experience of actors (in this case customers) in a (business) process [55]. This focus can resemble the evaluation of the experience of actors in a waste management process. We therefore make a cautious flip in the customer journey thinking that can unlock the systemic view on waste management processes (Fig. 2).

In essence, the flip entails changing the customers' perspective to the waste perspective. This means that the analytical ideas of a journey are now used to follow how the waste goes through the process of human-to-human handling and interacts with each of them at sequential points across the journey. The idea is that at these points of interaction, the analysis looks at different actors' intentions and behavior. This helps to establish a detailed overview of the waste management process by evaluating the interactions, relationships and arguments for actions and decisions.

Essentially, we define a Waste Journey as the process that waste goes through, across a sequence of events interaction points (made up of actions and decisions with regards to the handling of the waste) of actors that make up the waste management process. In the next part, the methodological components for this concept are synthesized with useful building blocks from the reviewed literature.

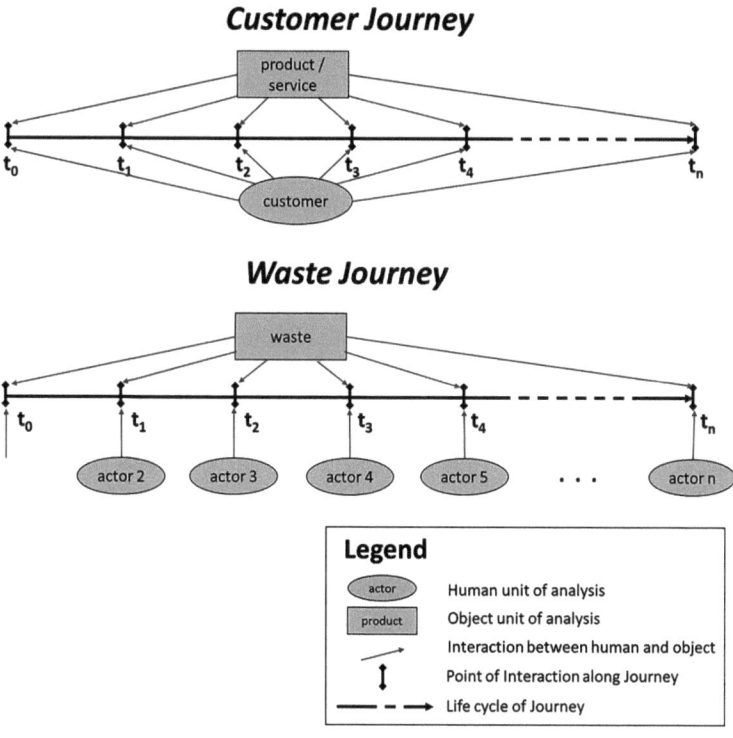

Fig. 2 Waste Journey conceptualization inspired by Customer Journey

3 Waste Journey Methodology

For a waste journey to work, choices need to be made regarding its set up, empirical data collection and analyses. First, we describe the setup of the Waste Journey concepts' goal and analytical components, as it is inspired by the customer journey. Next, we suggest the data collection, analysis, and visualization tools for implementing the Waste Journey method in real cases.

3.1 Waste Journey Setup

The goal of the waste journey is to identify barriers and opportunities for realizing inclusiveness and circularity in the waste management process. This implies an open research approach in which the waste is simply followed as it moves from actor to actor, starting at the point where resources have been discarded and enter the waste management process. By mapping the sequence of this process, the realization of the circularity and inclusiveness can be qualitatively analyzed. This is where the theory of planned behavior can help in identifying the underlying arguments for certain decisions and actions of the different actors can unveil experienced barriers and opportunities by these actors.

In the process, actors can act and decide in ways in which the waste process is performed in a circular way or not. For example, the waste can go to certain circular end-states across the journey and points where various actors interact with it. Therefore, the circularity of the waste journey can be defined by the end-state of the waste, like the preparation for reuse or incineration. The analysis observes the reasons for people to generate waste and how they will prepare themselves for this generation. Furthermore, people will take steps after the waste is generated and then the waste will go across different actors as it is picked up, processed, and distributed onto other locations, until a certain end-state. In the process the different actors can also interact with each other while handling the waste, for example, when waste is moved from hand to hand. Certain behavior will describe the actions that have led actors to behave in that way. The arguments behind this behavior can be clarified by using the theory of planned behavior, which can help to explain the potential origins of certain desired or undesired outcomes.

Figure 3 shows the graphical representation of the waste journey. In this scheme, each point of interaction represents a step in the waste management process, where one actor handles the waste and then leaves it or hands it over toward the next point of interaction. Within the point of interaction, the waste journey considers that these actions can be observed, and that the underlying planned behavior constructs (i.e. attitude, subjective norm, situational factor, knowledge, perceived behavioral control and intention) can be discerned. As the action or decision can help to understand the circular performance, the constructs of planned behavior help to understand the reasons of why these occur. Examples of these insights could be that an actor separates waste in a certain fashion because his neighbor does this too, or she has a positive attitude toward a certain decision.

Also, the theory of planned behavior can help to describe the inclusiveness of the waste management process. An example of this can be someone does not separate plastic waste, because there is no plastic container nearby. This way, it could for example be learned that some structural aspects, like accessibility of facilities, could be an underlying reason for certain circular performances not to be realized in the waste management process. This can then point to improvement suggestions.

A sensible method also has a clear demarcation of its system boundaries before it can be applied. Therefore, the waste journey has a clear start and end allowing the analysis of its circularity and inclusiveness. The start of the journey is when an actor decides that a resource is discarded as unusable. At this moment waste is generated. Then a varying number of steps could be taken by various actors in the journey that the waste undergoes. These steps can be circular and inclusive or the opposite of these. At some point the waste gets to an end of the journey. This happens when the waste has found a new stationary function (e.g. reused), dysfunction (e.g. at a junkyard) or in some other way eliminated (e.g. incineration).

The sequence of the data collection is chronological, i.e., starting at the closest actor to the start of the journey. This way the study can remain open to any direction that the waste is taken on its journey. This facilitates a detective type of study of the waste flow. It will also remain possible to prepare the study of a waste journey by sequencing the expected processing phases of the journey beforehand. The data can be collected at the points of interaction where an actor during an interview can describe the way they approach the waste action and decisions along the lines of the underlying constructs of the theory of planned behavior. This can be repeated for all actors until the end of the journey is reached.

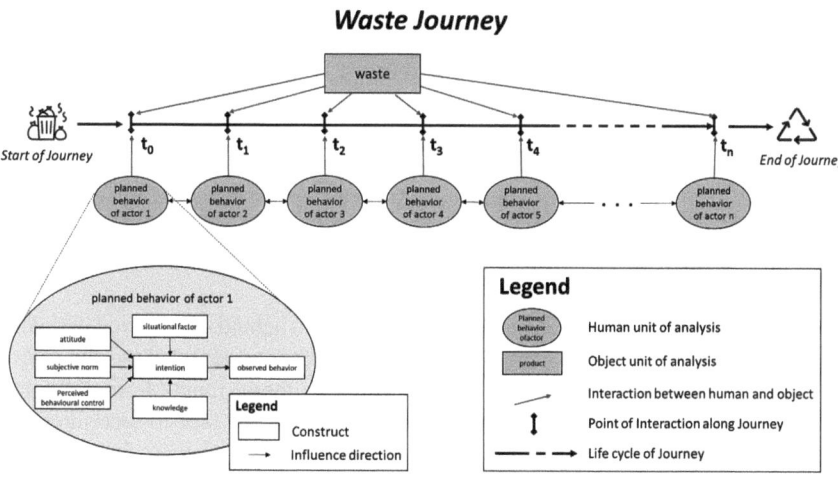

Fig. 3 Graphical representation of Waste Journey setup

An important sidenote about the data collection is also that waste at a certain point can be split into multiple follow-up waste items which receive their own continued journey. If this happens, then the approach faces a methodological choice to either continue with all journeys or with a prioritized one. Having a clear answer to such a query beforehand helps to avoid stalling the data collection.

The full data that can be used for further analysis in the end are the interviews about the constructs with the actors. The quality of the data set will depend also on the cohesive following of the actual journey of a waste item. This means that a physical type of waste can be used as actual flow to go to interview the actors that have handled it. This would, however, be quite restrictive. A more practical way would be to apply a snowballing technique in an interview to learn about the next actor that a certain focal waste item would be transferred to. The outcome of the full set of data can help generate an understanding of the waste as it has travelled and the why of it. This can lead to the detection of gaps and opportunities for intervention to achieve inclusiveness and circularity in the waste management process.

3.2 Data Collection

The waste journey is set up as a case study focusing on one waste stream. A frequently used method for data collection in case studies is interviews [26]. This helps to understand the perspective of a respondent and the how and why of that perspective. Interviews can be held both face to face or through an online platform, like Microsoft Teams. These can also be performed in various forms, like semi-structured, exploratory, or unstructured. The semi-structured approach is known for its flexible and versatile qualities during the conversation with a respondent [25]. This is an especially useful approach to gather perceptions and opinions from respondents, like on inclusiveness and circularity.

The interviews aim to reveal the waste journey as these are carried by the actors, who have been selected as the interviewees. Therefore, the interview questions target to map the process (sequence of decisions and actions) of the actor. Collecting a sequence like this is often helped by asking respondents to write the actions and decisions on cards, e.g. post-its, and then ask them to place these in front of them on a table. As decisions and actions are added on the surface, the chronological sequence of all these events can be constantly reflected on by the respondent to represent the process as accurately as possible.

The interviewer can then ask about the motivations behind each action and decision placed on the cards, allowing for inquiry into the reasoning behind the actor's behavior. This approach enables the collection of different inputs from the actors without needing to have them written down beforehand. A face-to-face interview provides the opportunity to also gather non-verbal communication with the respondents. This can be useful when interpreting whether a follow up question on their intentions, or other behavioral constructs could reveal additional insights form their perspective.

The interview protocol is made up of three parts. First, there are more generic questions which intend to obtain background information about the start of the waste journey and the self-described role and scope of experience of the waste journey of the actor. This is important to understand, because each actor forms a part of the waste journey but will probably not possess full knowledge of parts that occur before or after their involvement. Second, there are specific questions that one wants to ask with regards to the specific actor-role and phase that the individual performs. In this part it is useful to ask about the key arguments of the interviewees and their interactions with the direct actors before and after them in the journey. Regrading the waste flows, both quantitative and qualitative elements will be collected. For quantitative data, we focus on the amount and composition of generated waste flows. For qualitative data, we collect norms, value ideals, drivers, and routines of transactions.

3.3 Data Analysis and Visualization

The analysis of the data can be conducted in different ways, depending on the way that data was collected. First, a sequence of decisions and actions is mapped by an actor (which can be by post-it notes or on a Mural/Miro board canvas). The maps from individual actors can then be stacked up to reconstruct the collective waste journey. This can be visually represented through Draw.io or some other diagramming software application. Second, the interview texts can be coded along the planned behavior constructs (attitude, perceived behavior control, and so on) to help describe the arguments for actors' decisions and actions. The coding itself can be done by data-driven coding, i.e. coding with an open mind and no starting list of codes, or concept driven coding, i.e. coding along predefined categories or concepts [18]. A waste journey benefits from concept-driven coding because, if an actor's underlying motivations are already theorized under an umbrella like the theory of planned behavior, then concept-driven coding helps to gather evidence on these factors.

To enhance the understanding of the behavioral background behind waste flows the analysis should be presented in a narrative or textual manner and a graphical representation. A narrative representation entails the advantage of addressing the actor's underlying motivations and arguments. The graphical representation of a waste journey depicts waste flows from one actor to another. In each step actions of individual actors and the drivers behind these actions are provided. This representation gives a detailed overview of the actors' actions and the arguments they had for their actions but lacks details in all the steps in the waste management process. A swim lane diagram facilitates the visual insight by allowing a flowchart for the actions and decisions to unfold. Swim lane diagrams have this quality because it clearly presents which actor takes which action and the interactions between the actors (Lucidchart, n.d.). In addition, the swim lane diagram provides a degree of detail that helps in identifying potential bottlenecks (Lucidchart, n.d.).

4 Implementing the Waste Journey on Household Renovation Waste

In this part we report on the case of the HRW journey. First, we introduce the motivation of selecting HRW for our case study. Then we describe the context of the HRW in the Netherlands and the theoretical overview of the case-specific waste journey. Subsequently, we present and analyze two cases of HRW journeys. Finally, we describe patterns and revealing insights from these two cases.

4.1 Case Selection

For a proper empirical study on the applicability and academic novelty of the waste journey method, construction waste is a typical waste stream that has a lot of potential, because of its oftentimes job-based waste generation, and thereby organic manifestation of the eventual waste handling. For example, it is mostly generated through construction and demolition projects. This offers the opportunity to test the method to see how it helps to trace this waste handling and actors' behavior in it.

Construction waste in general is a critical waste category, which also performs quite well in terms of circularity in the Netherlands. The recycling rate of construction waste in the construction projects in the Netherlands is more than 95% [34]. In addition, quite some research is performed into this type of waste [36, 64, 67].

Within this major category, renovation waste is generally unknown regarding its degree of circularity. In this subcategory, single-house renovation projects are characterized by a high degree of decision-making due to the demolitions, which are more unpredictable for the actors involved. For example, there are typically small renovations of a bathroom, living room, or complete house for which the type of tiles, cabinets or other stuff inside the building can be quite hidden before waste is generated. The waste related to these projects is called Household Renovation Waste (HRW).

A standard waste management process can be understood in the following stages [7]: collection, transport, sorting, treatment, and final disposal or reprocessing. The customer journey informs that this overview misses a preparation and a generation stage. Therefore, as a matter of theoretical comprehensiveness, for the case it is decided to incorporate a preparation and generation stage in the standard waste management process.

Figure 4 represents the theoretical overview of the case-specific waste journey. The stages are used to check the journey completeness of the empirical study. It starts with the preparation, where waste could be generated, for what reasons, and inform on different routes. For example, a household during their renovation project could involve hiring a contractor to do the renovation for them. The different decisions made in this phase will have an impact on the remaining waste journey. Next, in the generation stage, the waste could either be generated by the household itself or

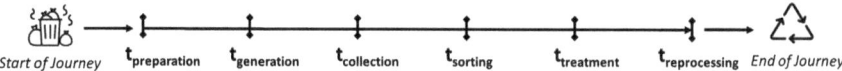

Fig. 4 HR waste journey stages derived from theory

by a professional organization. Different decisions could be made that influence the remainder of the waste journey, such as separating each tile with care, which makes them reusable, or by braking the tiles off the wall making them not reusable. In the collection stage, people can bring their waste to the municipal recycling center themselves, or the waste will be collected by a recycling/waste disposal company. In the sorting stage the waste is sorted and prepared for future purposes. In the treatment stage, the waste is processed for an end-purpose. In the disposal / reprocessing stage the waste is either disposed or reused, recycled, or repurposed.

4.2 Case Background and Contextual Assumptions

HRW can enter the waste journey through two optional routes. First, HRW can be brought to a municipal recycling center and will then be considered as Construction and Demolition waste or as bulky waste [45]. Another option for this waste to be managed is by using a container that is placed in front of the house. In this instance, there are containers in different sizes and for different materials, like wood, metals, bulky waste, and construction and demolition waste (Bouwbakkie.nl, n.d.). Insofar, neither of these options helps the separation of the HRW.

The actors in the HRW handling are represented in Table 1. It starts with citizens who own a home that they want to partly renovate. Once this is decided, citizens then make use of municipal recycling centers, contractors, and professional recycling companies for processing their waste. This makes the role of citizens in the HRW system quite big, and they need to be included in the system for it to function at its full potential.,

The HRW is expected to grow in the Netherlands in the period 2020–2030, based on a government goal aiming to insulate 1.5 million houses for realizing the energy efficiency in old homes [43]. Earlier estimates show that HRW after a home renovation project varies between 5 to 10 percent of the total construction and demolition waste [31, 35]. In these estimates, it is unclear if this includes only necessary retrofits or also esthetical renovations (Wang et al., 2020).

In the academic literature, it is hard to find studies that treat HRW as a separate waste stream, but as a part of the total construction and demolition waste. A few exceptions have studied renovation waste as a separate waste stream, and these note the importance to do so, because of the need to improve the reuse and recycling potential of HRW [8, 31, 35, 65].

Table 1 Overview of HRW actors

Actors	Roles
Household	Tend to generate waste by renovating something in their house. Can generate the waste themselves or hire a contractor. Can transport the waste to a municipal recycling center
Contractor	Generate HRW. Can do the transport, but not necessarily
Municipality	Set regulations for the waste management process. Supply the municipal recycling center
Waste disposal company	Offer containers for HRW collection and trans-ports them to the sorting facility. Can do the sorting itself but not necessarily
Waste sorting company	Treat the waste into different waste streams for the purpose of recycling. Can also do the collection of waste
Household	Tend to generate waste by renovating something in their house. Can generate the waste themselves or hire a contractor. Can transport the waste to a municipal recycling center
Contractor	Generate HRW. Can do the transport, but not necessarily

Among this small set of studies, two different definitions of HRW can be recognized. First, HRW is defined as waste that is related to "the modification or improvement of the residential building" [8, p. 1]. Second, HRW is defined as "waste that is related to the work needed in the building due to obsolescence or deterioration of some of its elements" [31, p. 392].

Ding et al. [8] defined six stages of HRW generation: layout transformation, installation engineering, mason engineering, carpentry engineering, paint engineering and related installation. Each of these stages all relate to a reason for a certain type of waste to be generated. Ding et al. [8] estimated that for their study about 75% of the waste was generated during the layout transformation stage, which mostly consisted of rough stony materials, like bricks, concrete and tiles.

Specific data about HRW can be collected using a variety of techniques, like site inspections, interviews, and literature studies [8, 68].

4.3 Mapping HR Waste Journey

Here we will report on the HR waste journey. In the Dutch context it was found that there are two different routes for citizens to handle HRW. Therefore, two separate waste journeys were mapped: direct municipal offering (WJ1), or on-site container (WJ2).

4.3.1 Waste Journey 1: Direct Municipal Offering

WJ1 started with a homeowner who decided on a renovation project himself in his house in Rotterdam. As the HRW was followed, it went to a recycling center and then to a sorting company for construction and demolition waste. An overview of interviews is offered in Table 2.

Figure 5 offers a visual representation of WJ1. It started with a homeowner (interviewee WJ1-1) who removed a wall to connect a living room with an old bedroom. He reasoned that it would be cheaper to do it himself. The homeowner created a plan from old pictures and asked a handyman for advice and quality check. After this was confirmed, the plan was executed, with first cutting the power and then a careful test. Then the wall was demolished entirely. The wall had a wooden frame with insulation material, closed with an oriented standard board and gypsum panels. This material was turned into smaller pieces and placed in bags for easier transport. The homeowner brought this to the recycling center himself since his municipality offered this as a free service. He used his own car and rented a trailer.

The recycling center inspected the waste but did not inspect his credentials as a citizen of the city. The operator at the recycling center (interviewee WJ1-2) said this should be done. After the waste inspection it was indicated where the waste should be brought on the premises. The homeowner regretted not separating the waste more thoroughly, as he put the wooden beams and panels together that had to be sorted in the wooden container and contaminated gypsum container. This separation was not checked by a supervisor at the recycling center (interviewee WJ1-1). In the background the recycling center representatives (interviewee WJ1-2, WJ1-3, and WJ1-4) elaborated that the check could have been done without the homeowner knowing about it.

Up until this point the homeowner was able to indicate the waste journey directly. Thereafter the waste flow became less clear, without direct accounts of the waste. This is where the accounts from the recycling center representatives provide more

Table 2 Overview interviewees of WJ1 about direct municipal offering

Actor	Representative interviewees	Action/role in WJ1-DMO
Homeowner	Interviewee WJ1-1	Remove a wall in a house in Rotterdam
Recycling center	Interviewee WJ1-2	Operate a municipal recycling center and trade waste
	Interviewee WJ1-3	Operates a commercial recycling center and trade waste
	Interviewee WJ1-4	Operate 12 recycling centers in Noord-Holland and trade waste
Construction waste sorting company	Interviewee WJ1-5	Sort construction and demolition (C&D) waste

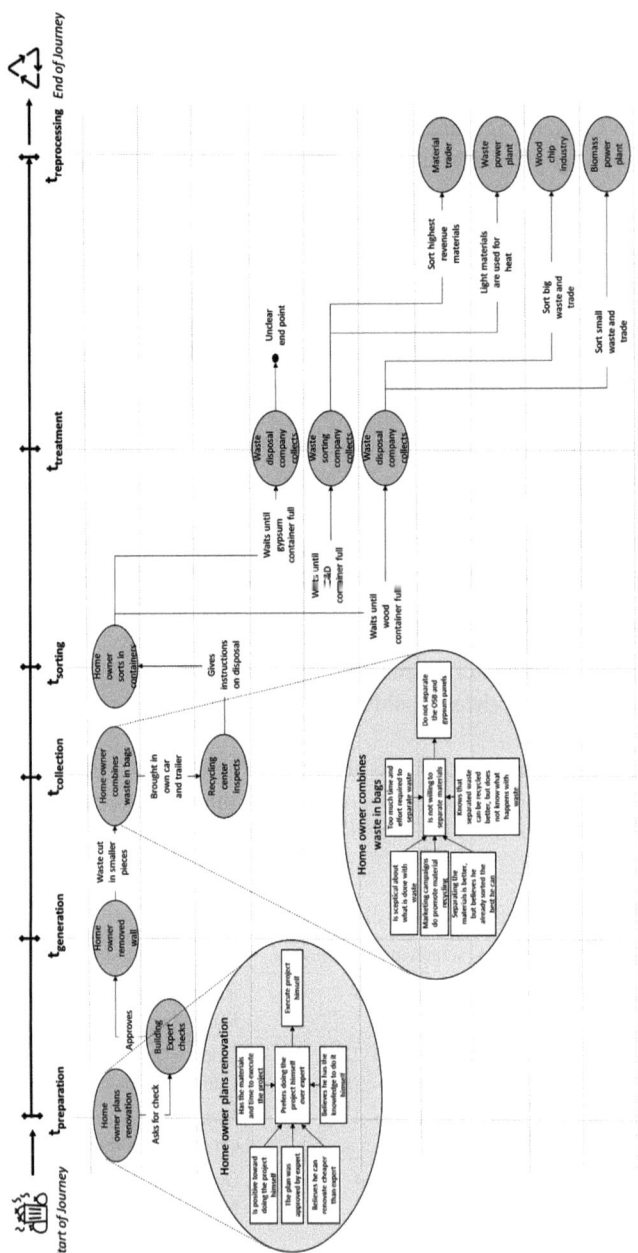

Fig. 5 Observed behaviors in Waste Journey 1

insights. These indicated that when the container is full, it will be collected, inspected, and weighed by a waste disposal company.

The wood container is sorted into different sizes. The bigger pieces can be used for woodchips in chipboards. The wood can also be compressed for pellets which will then be used in biomass power plants or for heating in houses.

The C&D container has a variety of materials due to its function as a residual bin. Also, this container is collected, inspected, and weighed by a waste disposal company. After the company accepts the container, it will pick the most valuable materials from its contents, like metals and chunks of wood. The remaining waste is then prepared to be sorted by a C&D waste sorting machine (interviewee WJ1-5). Once the waste arrives at the waste sorting company, it is specifically inspected for contamination and dangerous materials that can harm the machine, like gas cylinders, batteries, and video decks (which can strap around the machine).

Before the waste is inserted into the machine, a gripper picks up the biggest wooden and metal materials out of the waste pile. The waste sorting machine makes use of different techniques to automatically sort the waste. First, the waste is sorted into different sizes by using different sieves. Then high-caloric materials are sorted by wind shifters, i.e. automated machines which blow light materials away from stream. These materials are traded in different sizes, mostly as fuel for industries with a need for high, like the cement industries. After the wind shifters, the near infra-red (NIR) sorting machines separate wood materials out of the stream automatically. The different sized streams go into a handpicking station. Here, 8 to 12 people sort the waste by hand into containers underneath them, which are sorted for different purposes.

The rubble waste of wood comes out of these sorting processes and will be shredded. If the quality is good enough, then the shredded wood is traded as granulates or otherwise immobilized. Plastics are traded in the high caloric waste fraction or traded by plastic waste materials. The waste is traded as scrap metal.,

The contaminated gypsum container that comes from the recycling center was not mentioned during any of the other interviews. Therefore, it was not recorded in this waste journey what happened to this container.

In waste journey 1, we also aimed to capture the arguments of the observed behavior by actors in the journey. Given the challenges with the interviewees in this journey, we wanted to focus on two decisions from the homeowner that we could map in the best detail.

The first decision that the homeowner made was about performing the project himself. The main reason appeared to be the fact that doing the project himself would be cheaper than hiring an expert. He also had a positive attitude toward doing it himself, as he saw it as a fun challenge: "I found it a fun challenge, and it costs me a lot less money that way". Closest to the situational factor, the homeowner mentioned that he has the materials and time to do the renovation himself: "I started by drilling a small peephole in the wall to see......". The homeowner relied on the opinion for his plans from a known expert, which can be interpreted as a subjective norm.

The second decision that the homeowner made was about separating materials that were used from the wall. He decided that he did not want to separate the composite gypsum and OSB panels from the wall. This disables the circular potential of recycling both gypsum and OSB panels for reuse. Interestingly though, he was aware of this potential and the complications for non-separation: "I would suspect, it would be better to separate plaster and wood from each other as much as possible......but that would have taken me a lot of time and a lot of screwing". Yet, the homeowner was discouraged by the time and effort it would take to separate the two panel types. Also, he noted that gypsum would create a lot of dust: "And a lot of mess because that plaster, it's going to swirl everywhere". In addition, the homeowner was uncertain about what would happen to his waste thereafter: "I do not know what they are going to do with it. If it all gets flicked into the incinerator anyway." This created a certain skeptical attitude about sorting the waste: "I do not know how useful the work (waste separation) would be to me". A municipal marketing campaign did not convince him to separate it, because it did not tackle the concern about the time and effort: "If it all does end up going up in flames, I will have spent eight hours there screwing out screws for nothing and my whole house will be one big white dust mess".

4.3.2 Waste Journey 2: On-Site Container

WJ2 started with a contractor who completely renovated an apartment in Amsterdam. The contractor was willing to participate in the interviews. As HRW was followed we encountered a container broker, waste sorting, and a collection company. An overview of interviews is offered in Table 3.

Figure 6 offers a visual representation of WJ2. It starts when the contractor starts a complete renovation of an apartment in an old building complex in the center of a big city. The homeowner just bought the apartment according to the contractor (interviewee WJ2-1). The contractor and the client meet to align expectations and plan for the renovation work to commence.

The contractor prefers to completely strip the apartment and not reuse materials, because the quality of these cannot be guaranteed and preparing materials for reuse is costly and labor intensive. In addition, the contractor receives a margin on new materials, which incentivizes this procedure over reusing materials. For the renovation itself a permit was required, which included a reservation for a parking spot for

Table 3 Overview interviews of WJ2 about on-site container

Actors	Representative interviewees	Action/role in WJ2-OSC
Homeowner	No interview	Owns home to be renovated
Contractor	Interviewee WJ2-1	Completely renovate a house in Amsterdam
Waste sorting and collection company	Interviewee WJ2-2	Pick up containers, sort waste, and trade waste
Container broker	Interviewee WJ2-3	Arrange the container for the contractor

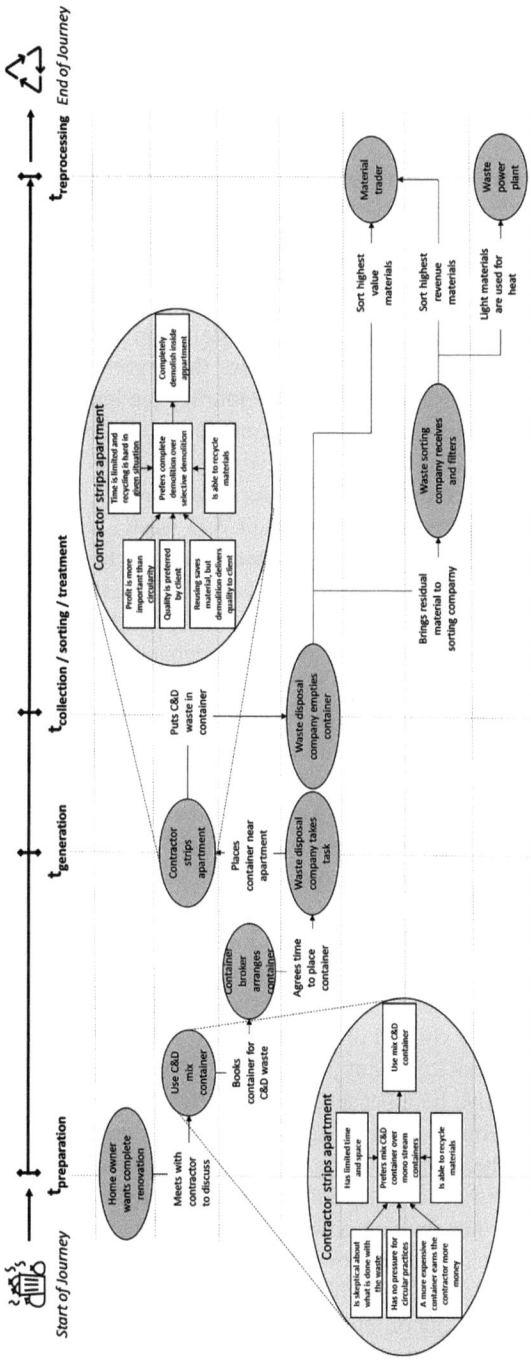

Fig. 6 Observed behaviors in Waste Journey 2

a container. This permit is requested and acquired by the homeowner at the municipality. After the permit was granted, the contractor contacted the container broker for arranging a container. The contractor prefers to use a mixed C&D Waste container, because sorting waste is more costly and laborious for their workers. The contractor also is skeptical about the waste sorting performed by the sorting companies.

The container broker guarantees the handling service for the container for the contractor and the waste disposal companies. The container broker finds communication between the parties important in his line of work, because of a general negative attitude and language barrier between the contractor and disposal companies. They communicate between the parties about the timely placement for the contractor and pick up for the waste disposal company. The waste disposal company is a separate company and has a waste sorting machine. The installation was not described.

In waste journey 2, we also aimed to capture the arguments of the observed behavior by actors in the journey. In this waste journey we had a contractor (interviewee WJ2-3) who could make multiple decisions that influenced the circular performance of the project. Therefore, we took the two key decisions from the contractor that we could map in the best detail for the observed behavior.

First, the contractor decided to not separate C&D waste, and instead opted for one on-site container for C&D mixed waste. One reason for this is also that sorting waste within the center of a big city is time consuming. Also, the contractor showed a skeptical attitude toward waste sorting: "I got a question about that the other day for a permit, some things are separated in there but to what extent is that reused. They say some things are recycled there. But I don't believe it that much". Furthermore, the contractor earns money on the price of the container, and even more importantly, the contractor takes a margin in the more expensive mixed C&D container than a wood container: "I would rather if we just look at the money, have one container that I can throw everything in than have a container that I can only throw wood in". Finally, the contractor feels no pressure from the client or other outside influences to focus on material circularity in the project.

Second, the contractor decided to completely demolish the interior of the apartment instead of applying a selective demolition: "Our preference is to take an apartment or a house/building completely empty, so that then we're not building on someone else's work". This decision was made to accomplish a higher quality end result for the client: "If you put plaster against it, you must plaster very thickly to get it straight again. That's just not possible, we don't do it like that". Selective demolition was unattractive here because preparing for reuse was considered quite laborious and costly: "You could reuse those battens that are often on the ceiling, but then you must take them all out in one piece, which will already be a challenge". Also, it was expected that the client would prefer to receive a high-quality project: "The customer expects that (buying new materials) and pays for it". In addition, in its attitude the contractor did acknowledge that circularity would be better, but that the business case is not right for this project: "If we can make more money by separating waste, and recycling or reusing most stuff then we will, yes". In the end, it is striking to see that the contractor decided to strip although he was aware and still unwilling to pursue the saving potential with selective demolition because of the clients unwillingness to pay otherwise "So, then it becomes unattractive to the customer very quickly, because then they must start paying an awful lot of money to have old materials back in there that in themselves are just as good as the new materials".

5 Discussion

In this section we discuss added value and implications of the proposed waste journey method based on the results of the two applied waste journey cases. In so doing, we first discuss the critical notes from the two HR waste journeys. Next, we lay out how the innovative ways of data collection and analytics have contributed to revealing these findings. Finally, we reflect more broadly on the meaning of these points for the waste journey method and its future use.

5.1 Critical Notes

The two waste journeys revealed a few remarkable findings. First, it showed some patterns regarding the non-transparency of the information of a waste journey. This manifested itself in different reasons for the start and end of the waste journeys. Close to the start of the journey some actors were observed as unwilling to participate in the research. For example, the homeowner in WJ2 did not want to be interviewed. Closer to the end of the journey, some actors did not want to openly speak about their activities or journey follow-ups. For example, in case of the recycling center in WJ1, there was an unwillingness to share specific follow-up partners in the journey because of certain commercial interests that were at play. These particular observations about the household renovation waste reveals how data inaccessibility in diligent and systematic data collection can actually qualitative lay bare the lack of inclusiveness in the waste management system because of hidden actions and motivations to general inquiry and potentially between actors.

Another core finding was related to the explicitly described circular behavior and underlying intent in the waste journeys. There were a few instances where the lack of circular actions was laid bare in the open where also the underlying planned behavior constructs helped in locating the specific reasons that prevented circular behavior from occurring. For example, in waste journey 2 the contractor clearly prioritized full demolition because he considered profit more important than circularity, despite his awareness of societal arguments in favor of the opposite. Another example, in the same light, was the shared motivation not to decide for circular actions, because of skepticism about the in-transparency of the waste treatment higher up in the waste journey.

Both findings lay bare a critical catch 22 about these waste journeys. First, the waste journeys were not regarded inclusive because of the unwillingness from actors at the start of the journey to share the reasons for their decisions due to a lack of understanding about what happens further along the journey. This all the while circular actions actually seem to be dislodged because of this exclusive attitude of actors between each other and in general to begin with.

5.2 Added Value of the Analytical Properties of Waste Journey Findings

The observations could be made as a result of analytical properties in the design of the waste journey methodology. First, the type of visualizations of Fig. 5 and 6 have been specifically designed to show the combination of action flows and underlying behavioral drivers. This was done in response to a lack of methods standalone not able to show both a clear disentangling of the total range of activities between the actors along the journey, and the complex motives for these actions and decisions. Oftentimes underlying motives can be shown in a narrative style, by events analysis and underlying constructs [63], but this generally lacks the ability to place these narrated actions and reasons into overall system of actors. On the contrary, flow-oriented diagrams, like Swimlane diagrams (Lucidchart, n.d.), are used to show this overall system, but fail to clearly outline the behavior between actors through the journey. This is where we conceptualized a combination of the Swimlane diagram, events analysis, and underlying constructs in Fig. 5 and 6 as a sort of relay diagrams. Here we define relay diagrams as the ability to explicate the interactions between actors through arrows, but separating them in lanes, as they shape the waste journey as if they run a relay race. In this way, a story board for the waste journey is visualized and explained with underlying behavior constructs.

Another methodological trait that helped uncover the findings in the waste journey was related to the potential of detecting underlying motivations for actions and the potential to identify the source of incentive propositions to curb the undesirable ones. For example, the homeowner in waste journey 1 did not want to separate the gypsum from the OSB panels because it was too much effort. The deeper reason for this appeared to be the skeptical attitude about what happens with waste further into its journey. This therefore revealed that a potential incentive to curb this behavior could be to be more transparent about the way in which waste is processed later into the journey. Also, in waste journey 2 the contractor preferred to completely demolish the interior of the apartment, instead of sorting waste on site. His motivation was both client-driven and financially driven, although he was aware of the circularity potential. This clearly showed the system-based incentives that play part into his decision, and the inevitable implications this had for the following journey.

5.3 Power of Waste Journey

The waste journey method was intended as a detective-like approach to follow waste from actor to actor, to understand the actual processes that specific waste undergo in a local setting. The set up and first application of this method on two HR waste journeys has uncovered a few notable contributions. First, it was found that currently no method is targeted to the empirical interplay between inclusiveness and circularity. Waste journeys have unlocked the observational capability in its systemic line of enquiry at flow and underlying actor behavior level to find relations between inclusiveness and circularity in the two cases. More specifically, it helped to observe that certain circular actions were not taken by actors at the start of the journey due to a lack of inclusiveness of what happens to their waste toward the end of the journey. In essence, this could help in finding the key barriers and opportunities for local waste management practices and inform local and regional authorities on how to curb undesired circularity effects with inclusive-triggering incentives.

A second contribution is the very fact that the setup itself allows to empirically collect data about actor behavior in a supply chain through a relay function. It thereby goes beyond the work of Pongpunpurt et al. [39] who focused on the applicability of the theory of planned behavior on the motivations of individual actors on waste-related decisions. A specific insight about the importance of this supply chain orientation was also unveiled in the set-up of the household renovation waste journeys. Specifically, it appeared that the preparation phase proposed by Buch et al. [7] was very crucial for this type of waste stream, because of the laborious process that demolition brings for the start of the HR waste journey. It was also empirically confirmed, due to the crucial decisions at the start that influences the journey in terms of who becomes involved and what types of underlying motivations are leading the next journey stages.

A final contribution lies in the versatility and openness of the waste journey method. As it tries to map waste flows by following the waste on its path through the waste management process, the snowballing technique makes it possible to identify the inclusiveness of actors about their role and decisions. In this way, the blank responses along the journey also become a potential indicator about the transparency of actions and decisions in the actual waste management process. Even more so, in cases the waste sorting company revealed that there are instances where the interviewed representative of the actor does not make the decisions, potentially revealing more complex internal organizational structures of why certain activities are taken. Also, the study did not particularly underscore any methodological limitation prohibiting it from following other waste streams.

6 Conclusions

In this paper the waste journey method was developed and tested in response to the question: *How to identify barriers and opportunities for realizing inclusiveness and circularity in a waste management process?* The waste journey method was shown to identify potential barriers and opportunities by looking at the actor interactions in the waste handling, their motivations for making decisions, overseeing its journey implications and the systemic mishaps that poor inclusion and circularity in a journey can unveil.

In a practical sense this study offers a method that could be of direct value to waste management practitioners and related policy officers in a city. The direct value would manifest itself in the identification of observed behavior of actors in the waste chain and their underlying motives for that behavior. It was shown through two cases on household renovation waste how the inclusion and circularity can interact between the actors in the waste chain, and how these hold each other hostage in undesirable behavior if not targeting the underlying reasons for stalling behavior. The novelty lies in both the unconventional approach of reconfiguring a design-based tool called the customer journey by flipping units of analyses and the introduction of the relay diagram to visualize the actor interactions and motives in a system process.

The waste journey method has a few observed limitations. First, the waste journey in its proposed form is mostly able to address single waste types and heavily contextualized (in terms of location and time) at a high-quality detailed level. There are reasonable expected modifications needed before it could be entertained for entire waste systems in a city or even comparability between two single waste types. Second, the waste journey in its proposed form still is a time-consuming approach to track and trace the waste. This is especially limiting when the researcher is not able to immediately access the relevant actors through the snowballing process. This could potentially be resolved through more closed and mass-oriented data enquiries like surveys. Thirdly, although the systemic approach allows to learn about the transparency of actors about their role in the waste chain, it still makes the journeys' continued enquiry dependent on a positive willingness for the next actor to share their actions, decisions and follow up partners. Potentially, more detailed and preplanned versions of a waste journey could mitigate this dependency.

References

1. Adama O (2012) Urban governance and spatial inequality in service delivery: A case study of solid waste management in Abuja, Nigeria. Was Manage Res 30(9):991–998. https://doi.org/10.1177/0734242X12454694
2. Ajzen I (1991) The theory of planned behavior. Organ Behav Hum Decis Process 50(2):179–211. https://doi.org/10.1016/0749-5978(91)90020-T

3. Ali M, Geng Y, Robins D, Cooper D, Roberts W, Vogtländer J (2019) Improvement of waste management practices in a fast expanding sub-megacity in Pakistan, on the basis of qualitative and quantitative indicators. Waste Manage 85:253–263. https://doi.org/10.1016/j.wasman.2018.12.030
4. Blasi S, Ganzaroli A, De Noni I (2022) Smartening sustainable development in cities: strengthening the theoretical linkage between smart cities and SDGs. Sustain Cities Soc 80:103793. https://doi.org/10.1016/j.scs.2022.103793
5. Boulding KE (2013) The economics of the coming spaceship earth. In: Environmental quality in a growing economy. RFF Press, pp 3–14
6. Bouwbakkie.nl (n.d.) Afvalcontainer huren in Nederland. https://www.bouwbakkie.nl. Accessed on 30 Aug 2022
7. Buch R, Marseille A, Williams M, Aggarwal R, Sharma A (2021) From waste pickers to producers: an inclusive circular economy solution through development of cooperatives in waste management. Sustainability (Switzerland) 13(16). https://doi.org/10.3390/su13168925
8. Ding Z, Gong W, Tam VWY, Illankoon IMCS (2019) Conceptual framework for renovation waste management based on renovation waste generation rates in residential buildings: An empirical study in China. J Clean Prod 228:284–293. https://doi.org/10.1016/j.jclepro.2019.04.153
9. Ellen MacArthur Foundation (n.d.) Featured circular economy examples: Cities. https://ellenmacarthurfoundation.org/circular-examples-collection-cities. Accessed on 30 Jun 2023
10. European Commission (2008) Directive 2008/98/EC of the European Parliament and of the Council of 19 November 2008 on waste and repealing certain Directives. https://eur-lex.europa.eu/legal-content/EN/TXT/?uri=celex%3A32008L0098
11. European Commission (2015) Closing the loop—An EU action plan for the circular economy. https://eur-lex.europa.eu/resource.html?uri=cellar:8a8ef5e8-99a0-11e5-b3b7-01aa75ed71a1.0012.02/DOC_2&format=PDF
12. European Commission (2020) Circular economy action plan: The EU's new circular action plan paves the way for a cleaner and more competitive Europe. https://environment.ec.europa.eu/strategy/circular-economy-action-plan_en
13. Følstad A, Kvale K (2018) Customer journeys: a systematic literature review. J Serv Theory Pract 28(2):196–227. https://doi.org/10.1108/JSTP-11-2014-0261
14. Gao L, Melero I, Sese FJ (2020) Multichannel integration along the customer journey: a systematic review and research agenda. Serv Ind J 40(15–16):1087–1118. https://doi.org/10.1080/02642069.2019.1652600
15. Gao X, Nakatani J, Zhang Q, Huang B, Wang T, Moriguchi Y (2020) Dynamic material flow and stock analysis of residential buildings by integrating rural–urban land transition: a case of Shanghai. J Clean Prod 253:119941. https://doi.org/10.1016/j.jclepro.2019.119941
16. Gerometta J, Häussermann H, Longo G (2005) Social innovation and civil society in urban governance: strategies for an inclusive city. Urban Studies 42(11):2007–2021. https://doi.org/10.1080/00420980500279851
17. Ghisellini P, Cialani C, Ulgiati S (2016) A review on circular economy: the expected transition to a balanced interplay of environmental and economic systems. J Clean Prod 114:11–32. https://doi.org/10.1016/j.jclepro.2015.09.007
18. Gibbs GR (2012). Thematic coding and categorizing. In Analyzing qualitative data. Qualitative Research Kit: Analyzing Qualitative Data, 38–55. https://doi.org/10.4135/9781849208574
19. Giovannini M, Huybrechts B (2017) How inclusive is inclusive recycling? Recyclers' perspectives on a cross-sector partnership in Santiago de Chile. Local Environ 22(12):1497–1509. https://doi.org/10.1080/13549839.2017.1363727
20. Gutberlet J (2015) More inclusive and cleaner cities with waste management coproduction: insights from participatory epistemologies and methods. Habitat Int 46:234–243. https://doi.org/10.1016/j.habitatint.2014.10.004
21. Hartmann C (2018) Waste picker livelihoods and inclusive neoliberal municipal solid waste management policies: the case of the La Chureca garbage dump site in Managua, Nicaragua. Waste Manage 71:565–577. https://doi.org/10.1016/j.wasman.2017.10.008

22. Hoornweg D, Bhada-Tata P (2012) What a waste : a global review of solid waste management. Urban development series;knowledge papers no. 15. © World Bank, Washington, DC. http://hdl.handle.net/10986/17388 License: CC BY 3.0 IGO
23. IPCC (2022) Summary for policymakers. Climate Change 2022: Impacts. Adaptation and Vulnerability 6:3–33. https://doi.org/10.1017/9781009325844.001
24. Izdebska O, Knieling J (2021). Citizen involvement in waste management and circular economy in cities: key elements for planning and implementation. Eur Spat Res Pol 27(2):115–129. https://doi.org/10.18778/1231-1952.27.2.08
25. Kallio H, Pietilä AM, Johnson M, Kangasniemi M (2016) Systematic methodological review: developing a framework for a qualitative semi-structured interview guide. J Adv Nurs 72(12):2954–2965. https://doi.org/10.1111/jan.13031
26. King N (2004) Using interviews in qualitative research. In: Cassel C, Sumen G (eds) Essential guide to qualitative methods in organizational research. Sage Publication Ltd, pp 11–22
27. Kirchherr J, Reike D, Hekkert M (2017) Conceptualizing the circular economy: An analysis of 114 definitions. Resour Conserv Recycl 127(September):221–232. https://doi.org/10.1016/j.resconrec.2017.09.005
28. Liu Z, Schraven D, de Jong M, Hertogh M (2023) The societal strength of transition: a critical review of the circular economy through the lens of inclusion. Int J Sust Dev Wor Ecol: 1–24. https://doi.org/10.1080/13504509.2023.2208547
29. Lucidchart (n.d.) What is a Swimlane diagram. https://www.lucidchart.com/pages/tutorial/swimlane-diagram. Accessed on 15 Jun 2023
30. Ma W, Hoppe T, de Jong M (2022) Policy accumulation in China: a longitudinal analysis of circular economy initiatives. Sustain Prod Consump 34:490–504
31. Marrero M, Rivero-Camacho C, Alba-Rodríguez MD (2020) What are we discarding during the life cycle of a building? Case studies of social housing in Andalusia, Spain. Waste Manage 102:391–403. https://doi.org/10.1016/j.wasman.2019.11.002
32. Mbah PO, Nzeadibe TC (2017) Inclusive municipal solid waste management policy in Nigeria: engaging the informal economy in post-2015 development agenda. Local Environ 22(2):203–224. https://doi.org/10.1080/13549839.2016.1188062
33. Merli R, Preziosi M, Acampora A (2018) How do scholars approach the circular economy? A systematic literature review. J Clean Prod 178:703–722. https://doi.org/10.1016/j.jclepro.2017.12.112
34. Ministry of Infrastructure and Environment & Ministry of Economic Affairs (2016) A circular economy in the Netherlands by 2050. https://www.government.nl/documents/policy-notes/2016/09/14/a-circulareconomy-in-the-netherlands-by-2050. Accessed on 20 Jun 2023
35. Mália M, De Brito J, Pinheiro MD, Bravo M (2013) Construction and demolition waste indicators. Waste Manage Res 31(3):241–255. https://doi.org/10.1177/0734242X12471707
36. Obaid AA, Rahman IA, Idan IJ, Nagapan S (2019) Construction waste and its distribution in Iraq: an ample review. Indian J Sci Technol 12(17):1–10. https://doi.org/10.17485/ijst/2019/v12i17/144627
37. Oguntoyinbo OO (2012) Informal waste management system in Nigeria and barriers to an inclusive modern waste management system: a review. Public Health 126(5):441–447. https://doi.org/10.1016/j.puhe.2012.01.030
38. Pearce DW, Turner RK (1989) Economics of natural resources and the environment. Johns Hopkins University Press, Baltimore
39. Pongpunpurt P, Muensitthiroj P, Pinitjitsamut P, Chuenchum P, Painmanakul P, Chawaloesphonsiya N, Poyai T (2022) Studying waste separation behaviors and environmental impacts toward sustainable solid waste management: a case study of Bang Chalong Housing, Samut Prakan, Thailand. Sustainability (Switzerland) 14(9). https://doi.org/10.3390/su14095040
40. Potting J, Hekkert M, Worrell E, Hanemaaijer A (2017) Circular economy: measuring innovation in the product chain—policy report. https://www.pbl.nl/sites/default/files/downloads/pbl-2016-circulareconomy-measuring-innovation-in-product-chains-2544.pdf
41. Rana J, Gaur L, Singh G, Awan U, Rasheed MI (2021) Reinforcing customer journey through artificial intelligence: a review and research agenda. Int J Emerg Mark. https://doi.org/10.1108/IJOEM-08-2021-1214

42. Rawal N (2008) Social inclusion and exclusion: a review. Dhaulagiri J Sociol Anthropol 2:161–180. https://doi.org/10.3126/dsaj.v2i0.1362
43. Rijksoverheid (2019) Klimaatakkoord—C Afspraken in sectoren C1 Gebouwde omgeving. https://www.klimaatakkoord.nl/gebouwdeomgeving/documenten/publicaties/2019/06/28/klimaatakkoord-hoofdstukgebouwde-omgeving. Accessed on 15 Aug 2022
44. Rijkswaterstaat (2021a) Beleidskader LAP3. https://lap3.nl/beleidskader/. Accessed on 15 Jun 2023
45. Rijkswaterstaat (2021b) LAP 3—Bijlage 3; Lijst van gebruikte termen, begrippen en definities. https://lap3.nl/beleidskader/deel-f-bijlagen/bijlage-3-termen/. Accessed on 15 Jun 2023
46. Sakamoto JL, de Souza S, Lima Cano N, Dionisio F, de Oliveira J, Rutkowski EW (2021) How much for an inclusive and solidary selective waste collection? A Brazilian study case. Loc Environ 26(8):985–1007. https://doi.org/10.1080/13549839.2021.1952965
47. Santana S, Thomas M, Morwitz VG (2020) The role of numbers in the customer journey. J Retail 96(1):138–154. https://doi.org/10.1016/j.jretai.2019.09.005
48. Santos S, Gonçalves HM (2021) The consumer decision journey: A literature review of the foundational models and theories and a future perspective. Technol Forecast Soc Chang 173(May):121117. https://doi.org/10.1016/j.techfore.2021.121117
49. Scheinberg A, Simpson M (2015) A tale of five cities: using recycling frameworks to analyse inclusive recycling performance. Waste Manage Res 33(11):975–985. https://doi.org/10.1177/0734242X15600050
50. Shavitt S, Barnes AJ (2020) Culture and the consumer journey. J Retail 96(1):40–54. https://doi.org/10.1016/j.jretai.2019.11.009
51. Steuer B (2021) Identifying effective institutions for China's circular economy: bottom-up evidence from waste management. Waste Manage Res 39(7):937–946. https://doi.org/10.1177/0734242X20972796
52. Steuer B, Ramusch R, Part F, Salhofer S (2017) Analysis of the value chain and network structure of informal waste recycling in Beijing, China. Resour Conserv Recycl 117:137–150
53. Terra L, Casais B (2021) Moments of truth in social commerce customer journey: a literature review. Springer International Publishing. https://doi.org/10.1007/978-3-030-76520-0_24
54. Tong X, Tao D (2016) The rise and fall of a "waste city" in the construction of an "urban circular economic system": the changing landscape of waste in Beijing. Resour Conserv Recycl 107:10–17. https://doi.org/10.1016/j.resconrec.2015.12.003
55. Towers A, Towers N (2022) Framing the customer journey: touch point categories and decision-making process stages. Int J Retail Distribut Manage 50(3):317–341. https://doi.org/10.1108/IJRDM-08-2020-0296
56. Tran HP, Schaubroeck T, Nguyen DQ, Ha VH, Huynh TH, Dewulf J (2018) Material flow analysis for management of waste TVs from households in urban areas of Vietnam. Resour Conserv Recycl 139:78–89. https://doi.org/10.1016/J.RESCONREC.2018.07.031
57. UN-Habitat (2010) Solid waste management in the world's cities: water and sanitation in the World's Cities 2010. London, UN-Habitat. https://unhabitat.org/solid-waste-management-in-the-worlds-cities-water-and-sanitation-in-the-worlds-cities-2010-2. Accessed on 15 Jun 2023
58. United Nations (n.d.) Social Inclusion | Poverty Eradication. https://www.un.org/development/desa/socialperspectiveondevelopment/issues/social-integration.html. Accessed on 20 Jun 2023
59. United Nations Department of Economic and Social Affairs. (2016). The report on the world social situation (2016) Leaving no one behind: the imperative of inclusive development. United Nations publication, New York
60. United Nations Statistics Division (2019) Goal 11—SDG indicators. https://unstats.un.org/sdgs/report/2019/Goal-11/. Accessed on 30 Jun 2023
61. Varnali K (2019) Understanding customer journey from the lenses of complexity theory. Serv Ind J 39(11–12):820–835. https://doi.org/10.1080/02642069.2018.1445725
62. Vasconcelos LT, Silva FZ, Ferreira FG, Martinho G, Pires A, Ferreira JC (2021) Collaborative process design for waste management: co-constructing strategies with stakeholders. Environ Develop Sustainab: 0123456789. https://doi.org/10.1007/s10668-021-01822-1

63. Van de Ven AH (2007) Variance and process models (Chapter 5). Engaged scholarship: a guide for organizational and social research. Oxford University Press, USA, pp 143–160
64. Villoria-Sáez P, Porras-Amores C, del Río Merino M (2020) Estimation of construction and demolition waste. In advances in construction and demolition waste recycling. Elsevier Ltd. https://doi.org/10.1016/b978-0-12-819055-5.00002-4
65. Wang J, Teng Y, Chen Z, Bai J, Niu Y, Duan H (2021) Assessment of carbon emissions of building interior decoration and renovation waste disposal in the fast-growing Greater Bay Area. China. Science of the Total Environment 798:149158. https://doi.org/10.1016/j.scitotenv.2021.149158
66. Wilson DC, Rodic L, Cowing MJ, Velis CA, Whiteman AD, Scheinberg A, Vilches R, Masterson D, Stretz J, Oelz B (2015) "Wasteaware" benchmark indicators for integrated sustainable waste management in cities. Waste Manage 35:329–342. https://doi.org/10.1016/j.wasman.2014.10.006
67. Yazdani M, Kabirifar K, Frimpong BE, Shariati M, Mirmozaffari M, Boskabadi A (2021) Improving construction and demolition waste collection service in an urban area using a simheuristic approach: a case study in Sydney, Australia. J Clea Prod 280:124138. https://doi.org/10.1016/j.jclepro.2020.124138
68. Zhang C, Hu M, Sprecher B, Yang X, Zhong X, Li C, Tukker A (2021) Recycling potential in building energy renovation: A prospective study of the Dutch residential building stock up to 2050. J Clean Prod 301:126835. https://doi.org/10.1016/j.jclepro.2021.126835
69. Zisopoulos FK, Steuer B, Abussafy R, Toboso-Chavero S, Liu Z, Tong X, Schraven D (2023) Informal recyclers as stakeholders in a circular economy. J Clea Prod: 137894. https://doi.org/10.1016/j.jclepro.2023.137894

Open Access This chapter is licensed under the terms of the Creative Commons Attribution 4.0 International License (http://creativecommons.org/licenses/by/4.0/), which permits use, sharing, adaptation, distribution and reproduction in any medium or format, as long as you give appropriate credit to the original author(s) and the source, provide a link to the Creative Commons license and indicate if changes were made.

The images or other third party material in this chapter are included in the chapter's Creative Commons license, unless indicated otherwise in a credit line to the material. If material is not included in the chapter's Creative Commons license and your intended use is not permitted by statutory regulation or exceeds the permitted use, you will need to obtain permission directly from the copyright holder.

Business Model Innovations in Post-Consumer Recycling in Urban China

Xin Tong

Abstract The innovation in business models for post-consumer recycling is booming in Chinese cities in recent years. This chapter illustrates the emerging business models for post-consumer recycling in urban China facilitated by the Internet. We identify three categories of emerging models: (1) community-based programs targeting the garbage sorting behavior of consumers for all household waste, (2) reverse logistic systems with automatic vending machines attached to traditional commercial chains, and (3) pure internet solutions to bridge the transactions between the consumers and recyclers. All these business models share the common characteristics that they use internet technology, which is aggressively promoted in China as "Internet+" by both government policies and venture capital investment. The various business models serve as the link between the firm and the system level and reflect the diverse possibilities for the future evolution of the recycling system in China. Five elements are key to the success of the business models, including convenience for consumers, traceability for producers, profitability for recyclers, hybridity for collection, and reliability of the information used by the public to address the various values pursued by different actors involved in the recycling chains. The results reveal the dilemmas facing each business model in balancing among all the elements and highlight the governance challenge of integrating the EPR scheme with the municipal waste management system.

Keywords Recycling · Extended producer responsibility (EPR) · Sustainable business model · Governance

1 Background

The Zero Waste movement has been embraced by local communities in many cities to regain sustainability and vitality through waste prevention and recycling [21]. The cross-scale efforts are devoted by various stakeholders in building the recycling

X. Tong (✉)
College of Urban and Environmental Sciences, Peking University, Beijing, China
e-mail: tongxin@urban.pku.edu.cn

common space in local communities [8]. The efforts include building pro-recycling environments in neighborhoods, planning recycling facilities for separated waste streams at the municipal level, and redefining the responsibility in waste management in the legal system. The innovation in business models for post-consumer recycling is booming in Chinese cities in recent years facilitated by internet technologies [20] as well as institutional change in waste regimes, such as the introduction of Extended Producer Responsibility (EPR), which provides an opportunity for fostering green innovation and new business models [17]. As an environmental policy approach intended, among other things, to create incentives for product innovation with lower environmental impacts throughout the lifecycle, EPR extends a producer's responsibility, physical and/or financial, for a product to its post-consumer stage, and shifts the cost of waste management from local government to consumers and producers. This policy strategy addresses not only the physical properties of a product, but also the related modes of consumption and production. These include promoting the provision of product functionality without relying on natural resource consumption, shifting towards product-service systems, and increasing interest in re-manufacturing activities within industries that produce and supply complex products [11]. As complex products often contain a large number of different materials with vastly dissimilar optical, mechanical, thermal, and electronic properties, it is essential to consider the recyclability at the very early stage of the product design. However, the practice of EPR in some sectors, for example, waste electrical, and electronic equipment (WEEE), shows that the simplified causal relations between regulation and producer's behavior assumed by the policy makers failed to address the complex interactions among various stakeholders [10]. Many factors could limit the effectiveness of EPR systems, including: commodity dynamics related to the volatile commodity prices that affect the value recovered from waste; volume dynamics due to the uncertainty in waste collection; competition dynamics related to the variations in the level of competition on EPR markets; regulatory dynamics caused by unexpected changes in future legislation; and design dynamics resulted from potential product design changes. Within EPR circle, these dynamics inevitably involved the government intervention for coordination [15].

This chapter presents our observation of an unexpected byproduct of EPR in China—the fostering of new business models for post-consumer recycling in cities. Emerging in several countries in the European Union (EU), EPR has been widely adopted in many countries including China. However, the local institutional settings for implementation in China differ from the original places where EPR developed. In the EU, a key factor in driving the adoption of EPR is that it shifts part of the cost of municipal waste management from local public expenditure to consumers and producers [1]. In most developing countries, however, post-consumer recycling is still thriving with an active informal recycling sector [13]. The experience in China could offer useful insights for the general South in upgrading the recycling sector by using new technologies in creating new business models.

2 Niches for the New Business Models

A business model describes how an organization may create, deliver, and capture value in various economic, social, and cultural contexts [16]. The institutional change could effectively stimulate the formation of new business model through redefining the rights and duties among stakeholders. Since 2012, China's regulation has mandated that the producers of certain categories of electronic products contribute to the government recycling funds based on their production volume according to the principle of EPR. The funds are used to provide subsidies to certified e-waste recyclers by the government. Although this system has been criticized as providing little incentive for design change or takeback actions by the producers [18], the subsidies have created market niches attracting investment and entrepreneurship devoting to recycling. Innovation in internet-based solutions for post-consumer recycling has exploded in the following years, most of which either benefited from the recycling funds, or aims to do so.

2.1 New Business Models Facilitated by Information Technology

We identified three major business models emerging in different cities in China: community-based recycling programs, automatic reverse vending machine chains, and pure Internet platforms. All these business models use Internet technology to track the flows of recyclables from the generation sources and provide incentives to users accordingly with various strategies in the relationship between online and offline activities.

2.1.1 Community-Based Recycling Program

This business model targets the garbage sorting behavior of consumers in residential communities. Examples include Ala in Shanghai (www.alahb.com), Xiangjiaopi in Beijing, Huishouge in Wuhan and Green Earth in Chengdu. Ala in Shanghai, the pioneer of this model, is a subsidiary company of Xinjinqiao, a state-owned certified e-waste recycler in Shanghai. Initially, Ala was established in order to find an efficient way to collect used household electronics from the residents in order to directly supply the recycling plant of Xinjinqiao. However, the company soon expanded its collections to include many types of recyclable goods in the communities. This model was followed by certified e-waste recyclers in other regions, such as Huaxin in Beijing (Xiangjiaopi) and GEM in Wuhan (Huishouge). In contrast, Green Earth (www.lvsediqiu.com) was established with funds from the Vantone Foundation, a charity fund established by a real estate company devoted to sustainable community programs in China. Green Earth intended to promote general garbage sorting in residential communities. They are also interested in whether e-waste collection could be a revenue source in future.

2.1.2 Automatic Reverse Vending Machine Chains

In this business model, a reverse logistic system with automatic vending machines is attached to traditional commercial chains. Examples include Aihuishou (www.aihuishou.com) in Shanghai and INCOM in Beijing. Aihuishou was established in 2011 with only an online trading platform for used information and communication technology (ICT) products. With three rounds of investment from venture capital since 2012, Aihuishou quickly developed a network of fashionable reverse vending machines in shopping malls and subway stations in big cities in China to collect used mobile phones. With support from the municipal government in Beijing, INCOM (incom.cc) initially copied a model from overseas to collect used drinking bottles with automatic reverse vending machines. Recently, they modified their machines to take back more items, such as used clothes, mobile phones and batteries.

2.1.3 Pure Internet Platform

This business model focuses on providing ICT solutions to bridge the transactions between the consumers and recyclers. Examples include Huishoubao (www.huishoubao.com) and Taolv365 (www.taolv365.com), both located in Shenzhen. Huishoubao cooperates with mobile phone producers to combine takeback of used products with sales of new ones. Taolv365 provides a trading platform for used mobile phones whose customers are mainly informal recyclers. Some large portal websites, such as Baidu, also provide their own takeback platform for trading of used products, but these systems are not so influential in the industry.

2.2 Key Elements of New Business Model

Extended producer responsibility is a market-based, life-cycle-oriented policy strategy to address the product-related environmental impacts. However, the static recycling targets at the end of life stage generally lead to inefficient market outcomes and weak incentives for prevention and green product design [3]. How to create and capture the value of waste reduction and recycling and share it among all the involved actors is a critical challenge in the design of the EPR scheme [14]. From 2009 to 2011 the national Home Appliances Replacement Scheme (HARS) (家电以旧换新计划) in China created an attractive market niche for the "old-for-new" model of producers to take back old products when selling new ones. In order to be qualified to receive the subsidy, the old products collected through this program had to be sent to the certified recyclers. After the termination of HARS, the government funds obtained from fees paid by producers according to EPR regulations were established as the long-term solution for WEEE recycling in China, which resume the market niche for new business models.

However, not all of the involved actors have been satisfied with the allocation of the funds raised through the producer fees along the recycling chain. On the one hand, the newly built recycling plants complain that the competition among certified recyclers in purchasing recyclable goods from the informal recyclers has squeezed their profits [4]. On the other hand, the formal channels established by the certified plants have difficulty competing with the informal collectors in efficiency and flexibility. The producers have little interest in being involved in the take-back activities after paying the recycling fees to the government funds. Knowledge gaps and value conflicts exist between various actors in forming an efficient collaboration.

Figure 1 shows a flow chart of information and material exchange among key actors under the EPR system. In order to create a closed-loop supply chain, an IT solution is used to provide a platform for information sharing along various actors. The information sharing has multiple effects: (1) It helps the consumers and the public authority to differentiate among recyclers according to their environmental protection standards. Currently, the newly-built formal recycling plants are equipped with the best available technologies and monitored according to very strict environmental protection standards. However, it is important to convey the information to the consumers for their decisions in waste disposal. (2) Theoretically, it can help producers to follow the status of their end-of-life products and inspire eco-design for new products according to the 3R principle (reduce, reuse and recycling). In 2017, the policy makers in China were discussing the possibility to remit a portion of the recycling fees if the producers can prove that they have taken responsibility for collecting used products on the market in the upcoming recasting of the China WEEE regulation. Thus, the tracking information in reverse logistics could be valuable for producers. (3) It can help the recycler to easily obtain information on the generation of recyclable goods and improve the value of collected goods on a wider and more transparent recycling market.

Various IT solutions have been developed respectively by different companies since 2011. And the focus of each system evolved towards different models as addressed in Sect. 3.1. During the implementation, several elements were identified as key for success according to the value to various involved actors along the recycling chain.

2.2.1 Convenience for Consumers

It has been argued that the take-back scheme should be convenient for consumers to return their products [1]. Generally, there are two ways for consumers to dispose of their old products. Before HARS, most of the consumers would sell their old products to the urban informal junk buyers who collected all categories of recyclable goods routinely in residential communities. But after HARS, more people prefer the

Fig. 1 IT solution as the center of the new business model

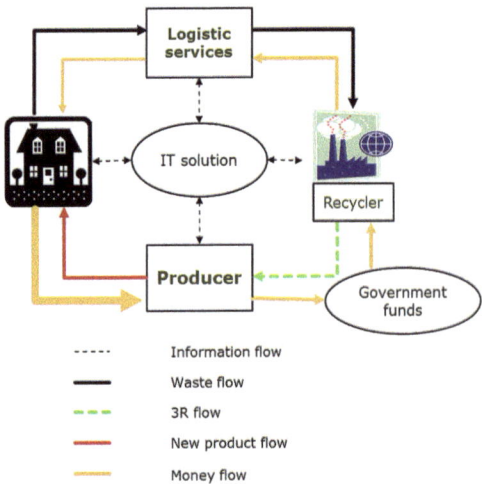

"old for new" model. In this model they receive discounts in purchasing new products, and retail stores arrange the collection of the discarded products. (In fact, they generally outsource the collection business to the informal junk buyers specializing in e-waste collection). The IT solution must employ one of the two channels as the main function. But none of the solutions can be integrated with both, for the on-line and off-line strategies are totally different. The biggest challenge to the community-based model is to build an image that allows it to distinguish itself from the informal junk-buyers. Thus, they need continuing presence in the community with routinized offline activities such as on-site promotion and door-to-door collection. The challenge to the old-for-new channels is finding a way to get the discounts for new product sales from the producers, either in traditional retail chains or on-line marketplaces.

2.2.2 Traceability for Producers

Producers need to be able to trace the flow of used products if they are to be able to benefit from EPR. As the response to the legal requirement for extended producer responsibility, producers increasingly accept that it is necessary to include reverse logistics in an integrated approach of closed-loop supply chain for their products [13]. The business value of returned products has been well-recognized [5]. The HARS established a standard process of "old for new" promotion for the electronics products in China. The producer can trace the old products they take back when selling new ones to verify the consumers eligible for discounts. When the returned products have high values on the secondhand market, the producers have great incentives to cooperate with the recyclers to explore this market niche. If the new product market and the secondhand one are not separated clearly, however, the producers' new product sales may be reduced. Thus, for example, Aihuishou provides the mobile phone producers a total solution of "old-for-new" promotion with traceability of all products being taken back so as to receive authorization from the producers to sell the secondhand products they collected from the end users.

2.2.3 Profitability for Recyclers

The recyclers see the potential to increase their profits through the new business models. In China, the government is trying to formalize the informal recycling sector. The formal sector receives subsidies from the government funds, while the informal recyclers can only rely on the market sales of the secondary materials recovered. The justification for such discrimination comes from the promise of the formal sector in their performance by increasing the stringency of environmental and social standards applied to recyclers. Thus, the recyclers have to accept very strict monitoring from local environmental agencies, which also increases the operational cost and decreases flexibility in the recycling process [22]. In order to enjoy the subsidy or authorization to conduct business, many formal recyclers are trying to build their own collection system. They expect that an information technology solution can provide an auditable record of their collection and recycling activities to differentiate themselves from the informal sector and thus justify the subsidy from the public funds for waste reduction and recycling with environmental sound technologies.

2.2.4 Hybridity for Collection

The existence of an extensive informal sector has been identified as one of the key challenges to develop a financially and environmentally sound recycling and disposal system for e-waste management in China [2]. Although the formal recyclers would like to exclude the informal collectors so as to control the collection channels, most of the formal recyclers found that it is much cheaper to rely on the supply from the informal collectors than to build their own collection channels. The hybridity—the mix of formal and informal enterprises—leaves space for various business models to fit in different channels, especially those for the secondhand markets for reuse and refurbishment [12]. However, in any business model we observed it is a challenge to balance the allocation of the value from collected goods between the formal and informal sectors. Some choose to exclude the informal sector and try to build a totally new image for the formal sector, while others are tempted to include the informal sector into the collection network. However, the rigid standards of the subsidy from the government funds generally reduce the profitability of recycling activities in the informal sector, which relies on flexible approaches to both waste conversions from material recovery and refurbishment/reuse according to the condition of products and market demands.

2.2.5 Transparency for Public

The expenditure of public funds on recycling requires transparency and reliability of information provided to the public [9]. The government needs to make sure that the certified recycling plants that receive subsidies recycle the e-waste in a proper way. The monitoring and traceability of the subsidy system was designed for the formal recycling system. However, wherever the formal system has to interact with the informal system, there are conflicts on the definition of the proper way to recycle. The complexity of transaction and material flows makes it inevitable that the formal system will have many open loops competing with or complemented by the informal sector. The open access to information is critical to public oversight of recycling activities.

2.3 *Diversification of Business Models*

According to the five elements, various new business models are emerging in different cities. Generally, the use of internet technology improves the traceability and reliability by tracking the waste flows from generation households to recycling sites, and incentives can be directly given to households based on the weight of recyclables they submit. In all business models, the off-line reverse logistics is essential in the competition among the companies which provide the transaction platform for recycling businesses. From this perspective, diverse paths are evolving among different models.

The first branch was upgrading the door-to-door collection process of the former street peddlers, who use the IT solutions to provide incentives to households participating in the community recycling program. The households can order the collection service at home through smart phone applications. These are the service-oriented business models, which have been adopted by Huge in Hangzhou and Aifenlei in Beijing. The recycling companies receive financial supports from the local governments to provide the door-to-door collection services in the community. The difference between the new business model and the traditional informal recyclers is that the role of the collector is not only to do the collection work, but also to provide face-to-face guidance to the residents to improve the quality of waste classification. And all recyclable waste generated within a household will be collected, including cheap materials, which will be otherwise sent to landfill or incinerators, such as glasses, foamed plastics, and so on, for the revenue from recycling these materials cannot cover the cost for collection. In 2022, Aifenlei has served 1009 communities in 6 districts in Beijing, covering more than 0.7 million households, and collected more than 110 thousand tons of recyclables. Meanwhile, Huge has served several districts in Zhejiang province, building more than 400 recycling stations, and serving more than 2,000 residential communities with more than 1.3 million registered users.

The second branch follows the paths towards intelligent facilities, focusing on the development of hardware facilities with improved intelligent technologies in order to save labor costs for the door-to-door services of collectors. Thus, the market value of the recyclables could be used to provide economic incentives to the participating households, and not relying on the financial support from the local government. The typical firms are Aobag from Chengdu and Little Yellow Dog from Dongguan, who received financial supports from the venture capital market in 2016–2018. Aobag set up the self-service drop site in the community, which is an unmanned space of about 20 square meters, open 24 h every day of the week to support members to drop their sorted recyclables. The recyclables dropped by the residents will be sorted into 14 categories, and the rewards will be deposited to the resident's accounts accordingly. By shortening the business chains and saving labors, the company can provide competitive compensation to their registered users. Little Yellow Dog integrated the waste sorting into an intelligent box, and quickly installed them in dozens of cities throughout the country. However, the quick expansion broke the capital chain, the investor went bankrupt in 2019.

Conflicts exist between the convenience to consumers and the profitability for recycling for both formal and informal sectors. The provision of off-line services, such as on-call service to pick up the waste products at home and recycling stations close to the users, is positive for the convenience to consumers in the community-based cycling program, but detracts from the profitability of recyclers. Especially for the formal recycling plants, economies of scale are significant for efficiency in operation. However, the automatic reverse vending machine chains could achieve both convenience to the consumers and the profitability for recyclers, but are only feasible for standardized products, such as drink containers or mobile phones.

Different strategies exist among these business models as to whether the informal sector is included in their collection system. The inclusion of the informal sector can raise the efficiency and flexibility of collection, but increase the complexity in monitoring the material flows and transactions.

In general, the tension between different elements in the business models raised the issue to balance the efficiency of recycling business and the social/environmental objects pursued at the system level. The compromises reflect the governance challenge between the EPR schemes and municipal waste management systems noted in the research literature [7].

3 The Impacts of Governance Structure on Business Models

EPR presumes a simple producer-driven commodity chain in the lifecycle of a complex product. However, complex interactions among all stakeholders along product chains challenged such a static view [6]. In practice, EPR provides an approach to waste issues that leads to a shift of authority from a local public service to activities entailing cross scale governance structures. Such a structure can match the scope of production networks of complex products making emergence of new governance structure possible. The differences in on-line/off-line strategies in the various business models reflect the variety of governance structures among the key stakeholders in post-consumer recycling in China. This, in turn, reflects the key conflicts between the logic of efficient close-loop supply chains anticipated by the EPR scheme and the traditional municipal waste management system operated by local public authority. In this section, we discuss the impact of governance structure on the emerging business models along two dimensions: (1) how the business models vertically integrate the production and consumption chain, including the upstream producers and downstream recyclers (either formal or informal); and (2) how the business models horizontally coordinate with the broader urban waste management system, such as garbage sorting in communities and reverse logistic systems that divert the recyclables from the municipal waste stream.

3.1 Vertical Integrations with the Product Chains

To benefit from the government funds for WEEE recycling, the new business models seek to forge a niche by bridging production and consumption/recycling which were generally separate before the introduction of EPR. As mentioned above, the consumer, producer, and recycler have various roles and objectives in participating in recycling activities. The governmental recycling fund created a value for recycling by collecting a small fee on each new product from producers and injecting the value back into the commodity chains by providing subsidies to the certified recycling plants. Several proposals have been widely discussed among policy makers and the industry to include the producers in take-back and recycling program, so as to materialize the incentive for eco-design from the beginning of new product development as the EPR scheme had expected. However, the complexity and conflicts in definition of eco-design prevent a universally accepted evaluation framework that can be used as a basis for differential fee system according to the quality of the producer's eco-design.

The practical motivation for producers to be involved in the takeback activities comes from sales of new products in an increasingly saturated market. Thus, the largest opportunity for value capture in the new business model comes from the discounts by producers in "old for new" sales promotion because it increases sales of new products to participating households. As more and more producers have included these "old for new" discounts as a necessary part of their new product marketing cost, the fee for the government recycling funds has become trivial in the total marketing cost. For example, one refrigerator producer we interviewed provided a discount of about 10% for consumers in old-for-new transactions which amounted to RMB 150-300 (USD 22-44) for each unit, much higher than the recycling fund rate at RMB 12 (USD 1.76).

Compared to large consumer electronics which have thin margins for refurbishment or recycling, used mobile phone recycling has attracted the most attention. Even though mobile phones had not been included in the catalogue of China WEEE regulation until 2015, business in the collection of used mobile phones has been booming in recent years due to both the expectation of the expansion of the catalogue and the obvious profitability in selling secondhand mobile phones. Taolv365 is one of the frontiers in providing a transaction platform for the sale of used mobile phones, established in 2009. As a pure internet platform, Taolv365 provides information on supply of and demand for different modules of used products. The users of Taolv365 are mainly informal recyclers. Recyclers from Shenzhen collect the discarded mobile phones through varied channels. They mail the goods in bulk to Taolv 365 in Shenzhen. Workers in Taolv365 check the products one by one, then classify them according to the components and quality required by the buyers. The buyers are also informal recyclers doing refurbishment or disassembly. There is a vast network of informal recycling around Shenzhen highly specialized in the division of labor along the recycling chains. The functional components taken apart from the used products are classified and sold on the informal market for repair, refurbishment, or other usage such as toy production. Taolv365 provides a bidding platform for sellers and buyers. The price index generated from the transactions on Taolv365 has been used in other IT solutions for pricing used products.

The emerging of two companies, Aihuishou and Huishoubao, exhibited different strategies in alignment with the EPR system as pursuing cooperation with producers, and defeated Taolv365 after 2018.

Based in Shenzhen, the largest mobile phone manufacturing region in the world, Huishoubao designed its business model to closely fit the "old for new" promotions of producers, actively involved in their partners' marketing promotions both online and offline. And Huishoubao chose to sell the used mobile phones collected from the consumers on secondhand markets by themselves, and to send the rest of the collected phones that were not of sufficient quality for reuse to the certified recycling plants. This model was welcomed by the major producers, because it provides very good traceability of the flows of used products. Counterfeit products have been embarrassing the brand producers in Shenzhen for years. Huishoubao promised not to refurbish the products they collected or remove any components, so as to prevent the outflow of components into the market for counterfeit production. This model provides a possible closed loop within the formal sector, from production, to consumption, then back to Huishoubao for reuse, and finally into the certified recycling plants for material recovery. It is a complement for "fast fashion" consumption in consumer electronics, which encourages consumers to easily move to the next generation product.

In contrast, Aihuishou, as a strategic partner with Xinjinqiao in Shanghai, focused its business on a transaction platform between the consumer and recyclers. It also provides the consumers with coupons for discounts in buying new products. The discounts, however, come from payments arising from a bidding process among recyclers based on the value on the secondhand market. This model is open to the informal recyclers, and cannot be used to trace the material flows after sale to the recyclers. With the support from venture capital, Aihuishou has invested in building extensive chains of automatic reverse vending machines in traditional retail centers in big cities, so as to compete with Huishoubao in attracting higher income consumers to bring back the most valuable used products.

Since mobile phones became eligible for government recycling funds in 2015, significant changes in business environment have been widely expected within the industry. Generally, the new business environment will probably favor the model of Huishoubao and Aihuishou, leaving decreasing margins for the informal refurbishment and disassembly activities prevalent in Shenzhen, which threatens the profitability of Taolv365.

3.2 Horizontal Relations with the Urban Waste Management System

The new business models for used mobile phone recycling represent possible solutions for post-consumer recycling of the "fast fashion" consumption products in which the cost of "old for new" is included in producer's new product promotion. The comparatively high value for second hand mobile phones and easy transportation of the products make this business model economically feasible. However, the business model is quite different from taking back the large used consumer electronic products, such as televisions, washing machines, refrigerators, air conditioners, and personal computers, which have been included the catalogue of China WEEE regulations for years. Although "old for new" has been a prevalent strategy for new product promotion of these products, it is the retail chains of electronic products that dominate the sales process. They generally outsource the take-back business to the informal recyclers. Since the end of the HARS, the certified recycling companies largely rely on informal channels to collect the waste products. The subsidy either from the HARS or from government recycling funds sets a floor price for the waste products. If the informal recyclers can find a better price than that given by the certified recyclers, the used products will not flow into the certified recycling plant, which creates environmental risks related to inappropriate recycling.

In order to improve the traceability of the take-back flows as well as increase the profitability of the certified recycling plants, some certified recyclers have been tempted to establish their collection system directly from the waste generation sources—the households in residential community. In contrast to the vertical specialization focusing on specific waste products such as mobile phones, this community-based model is horizontally integrated to encompass all sorts of recyclable goods, and closely interacts with the urban waste management system.

By establishing partnerships with local government, these companies actively engage in promotion of garbage sorting in communities. The growth of municipal solid waste has become a pressing environmental challenge in many cities in China with increasing burden on local public expenditure and NIMBY-ism related to the construction of waste disposal facilities [23]. Since 2000, various programs to promote garbage sorting and an urban circular economy have been initiated in different cities, either as national demonstration projects, or as bottom-up grass-roots experimental actions [19]. However, most of these efforts have been difficult to maintain. On the other hand, the informal recycling sector has been booming, and resulted in the prevalence of waste villages around many big cities. The government expected that the new business model could help the transformation of urban waste management systems.

With support from both the certified recycling plant and municipal government in Shanghai, Ala created a business model using IT technology to trace the garbage sorting behavior of households and provide incentives accordingly. They designed various activities on-site in the communities and broadcasts on public media to spread the knowledge about recycling to the public. Economic incentives were used as complement to the education. With continuous efforts lasting for several years, Ala has become widely known among Shanghai residents. However, their contribution of collections to the certified recycling plant was still very low, less than 10% of the recycling capacity of Xinjinqiao. The conflicts between the pursuit of economies of scale by the certified recycling company and the expectation on an all-inclusive solution to community recycling by the local government, prevented the certified recycling company from continued use of this business model.

A similar business model was adopted by other companies, such as Green Earth in Chengdu which is focusing on community recycling promotion, and Sound, the leading company in solid waste treatment in China. They are trying to integrate resource recycling and solid waste disposal into an all-inclusive solution for the local government with reduced burden on public expenditure and enhanced environmental performance.

The existence of the informal recycling sector, however, complicates the situation. All the three companies mentioned above have excluded informal recyclers from their collection system. They compete with the informal sector directly in the community by distinguishing themselves from the traditional urban junk-buyers. One exception is Huishouge in Wuhan. With door-to-door collection services in the community, Huishouge tried to open the platform to the urban junk buyers. They signed contracts with the junk buyers and provided information about the demands for collection of recyclable goods by household, then a nearby junk buyer would go to collect the goods from door to door. With the support from GEM, another leading solid waste treatment company headquartered in Shenzhen with many certified e-waste recycling plants in different cities, Huishouge operated for more than one year. However, just as with Ala, it contributed limited amounts of recyclable goods to the certified recycling plants of GEM.

In our ongoing experimental program in Beijing, we find that the inclusion of informal recyclers into the collection system challenged the local governance structure in waste management. Traditionally, the urban–rural segregation resulted in the division between the capital-intensive waste disposal system (the formal sector) and labor-intensive resources recovery (the informal sector). In order to include the informal recyclers in the community recycling program, the urban–rural segregation has to be ended. What is more, the junk buyers have quite a different role from the companies with new business model. Every junk buyer in the city is an individual entrepreneur, making every effort to improve the value of his goods. For example, one junk buyer we interviewed would be willing to bike for 3 h for the extra 10% for each kilogram of his goods. The most important value to them is the market price of the recyclable goods they collect. However, to the companies using new business models, their value arises from their image to the consumers. They devote considerable effort to creatively attract the consumers' attention and to maintain the consumers' participation in their programs. With the financial support either from the subsidies via recycling funds, or from the local public funds for waste reduction projects, the value of the recovered materials only contributes a fraction of their revenue.

From the perspective of the public interest, waste reduction and resource conservation has increasing value. The IT solutions used in the new business models make it feasible to track the volume and quality of the sorting process as a reliable measurement of waste reduction and resource recovery. However, it requires the governance structure to be flexible enough to bridge the formal and informal sector, so the value captured from the waste reduction and recycling can be shared among the stakeholders.

4 Conclusion

The EPR studies have revealed the governance challenge in urban waste management to capture the value from waste reduction and recycling, and sharing among all stakeholders in a complex and dynamic product system. Existing studies focused on the induced change in product design or business model at firm level, but few studies have examined the systematic change that the new business models could trigger in the recycling sector, especially in cities of the developing countries.

The change of institutional settings based on EPR principle in China provided an excellent empirical case to study how new business models could emerge as response to the government interventions. The introduction of EPR for WEEE in China allowed for innovation and the creation of new business models in the recycling sector and has triggered extensive changes in the business relations in the waste/recycling sector. It took an unexpected form in practice: not the producer doing the take-back themselves, but creating opportunities for various new business models to build links between the recycling and production/consumption regimes.

The most important technological change enabling the new business models is the use of ICT technology to bridge the knowledge gap among stakeholders including producers, recyclers, consumers, and the governments. This contributed to improve the transparency of EPR systems in the complex and fragmented product chains, and effectively respond to the market dynamics. Although, most of the new business models are heavily relying on the subsidies from the government at present, they show the possibility to incorporate the informal sector in developing countries, which has been emphasized profoundly in literature on WEEE in recent years.

Furthermore, this research reveals the diverse ways that the new business models connect the EPR system to the production network and the urban waste management system. On the one hand, the "old for new" business model, targeting secondhand products with high value and taking full advantage of on-line transactions, is favored by the venture capital investors. This is complementary to the "fast fashion" consumption of current consumer electronics industry. On the other hand, the community-based recycling program is closer to the local government's expectation of an "all-inclusive" solution to the waste management from bottom up. However, gaps existed between the economy of scale pursued by the certified recycling plants and the variety of waste reduction strategies in local community. This is one of the key problems in EPR studies—balancing the innovation-oriented policy for industry and the efficiency-oriented operation in waste management.

The case studies on various business models in China demonstrate the diversity of market niches and the responses in strategies of the entrepreneurs. The differences come from the conflicts and compromises in the values of different stakeholders reflected in the five elements we identified in the field investigation. In the evaluation of the performance of the major business models for post-consumer recycling emerging in China cities, we find that each business model has its own approach to balance between the efficiency of recycling and the broader social/environmental targets that the EPR scheme is intended to achieve. Therefore, there is not just one business model, but a variety of models to fit different institutional settings either within the vertical production chains at the macro level, or across the horizontal relations with the waste management at the local level. Therefore, EPR could open a gate to new business opportunities, beyond closing the loop of material flows within the product chains.

References

1. Cahill R, Grimes SM, Wilson DC (2011) Review Article: extended producer responsibility for packaging wastes and WEEE - a comparison of implementation and the role of local authorities across Europe. Waste Manage Res 29:455–479
2. Chi X, Streicher-Porte M, Wang MYL, Reuter MA (2011) Informal electronic waste recycling: a sector review with special focus on China. Waste Manage 31:731–742
3. Dubois M (2012) Extended producer responsibility for consumer waste: the gap between economic theory and implementation. Waste Manage Res 30:36–42
4. Gu Y, Wu Y, Xu M, Wang H, Zuo T (2016) The stability and profitability of the informal WEEE collector in developing countries: A case study of China. Resour Conserv Recycl 107:18–26
5. Guide D, Van Wassenhove L (2009) The evolution of closed-loop supply chain research. Oper Res 57:10–18
6. Hafkesbrink J (2007) Transition management in the electronics industry innovation system: systems innovation towards sustainability needs a new governance portfolio. In: Lehmann-Waffenschmidt M (ed) Innovations towards sustainability: conditions and consequences. Physica, Heidelberg, pp 55–86
7. Hickle GT (2014) An examination of governance within extended producer responsibility policy regimes in North America. Resour Conserv Recycl 92:55–65

8. Hutner P, Thorenz A, Tuma A (2017) Waste prevention in communities: a comprehensive survey analyzing status quo, potentials, barriers and measures. J Clean Prod 141:837–851
9. Kissling R, Coughlan D, Fitzpatrick C, Boeni H, Luepschen C, Andrew S, Dickenson J (2013) Success factors and barriers in re-use of electrical and electronic equipment. Resour Conserv Recycl 80:21–31
10. Lauridsen EH, Jørgensen U (2010) Sustainable transition of electronic products through waste policy. Res Policy 39:486–494
11. Lindhqvist T (2000) Extended producer responsibility in cleaner production. IIIEE, Lund, Lund University IIIEE Dissertations 2000:2
12. Liu H, Lei M, Deng H, Keong Leong G, Huang T (2016) A dual channel, quality-based price competition model for the WEEE recycling market with government subsidy. Omega 59:290–302
13. Manomaivibool P (2009) Extended producer responsibility in a non-OECD context: the management of waste electrical and electronic equipment in India. Resour Conserv Recycl 53:136–144
14. Massarutto A (2014) The long and winding road to resource efficiency—an interdisciplinary perspective on extended producer responsibility. Resour Conserv Recycl 85:11–21
15. OECD (2016) Extended producer responsibility: updated guidance for efficient waste management. OECD Publishing, Paris
16. Osterwalder A, Pigneur Y (2010) Business model generation: a handbook for visionaries, game changers, and challengers. John Wiley and Sons, New Jersey
17. Rossem CV, Tojo N, Lindhqvist T (2006) Extended producer responsibility: an examination of its impact on innovation and greening products. www.greenpeace.org/international/PageFiles/24472/epr.pdf. Greenpeace International, Friends of the Earth and the European Environmental Bureau (EEB)
18. Tong X, Yan L (2013) From legal transplants to sustainable transition. J Ind Ecol 17:199–212
19. Wang J, Han L, Li S (2008) The collection system for residential recyclables in communities in Haidian District, Beijing: a possible approach for China recycling. Waste Manage 28:1672–1680
20. Xue Y, Wen Z, Bressers H, Ai N (2019) Can intelligent collection integrate informal sector for urban resource recycling in China? J Clean Prod 208:307–315
21. Zaman AU, Lehmann S (2013) The zero waste index: a performance measurement tool for waste management systems in a 'zero waste city.' J Clean Prod 50:123–132
22. Zeng X, Duan H, Wang F, Li J (2017) Examining environmental management of e-waste: China's experience and lessons. Renew Sustain Energy Rev 72:1076–1082
23. Zhang DQ, Tan SK, Gersberg RM (2010) Municipal solid waste management in China: status, problems and challenges. J Environ Manage 91:1623–1633

Open Access This chapter is licensed under the terms of the Creative Commons Attribution 4.0 International License (http://creativecommons.org/licenses/by/4.0/), which permits use, sharing, adaptation, distribution and reproduction in any medium or format, as long as you give appropriate credit to the original author(s) and the source, provide a link to the Creative Commons license and indicate if changes were made.

The images or other third party material in this chapter are included in the chapter's Creative Commons license, unless indicated otherwise in a credit line to the material. If material is not included in the chapter's Creative Commons license and your intended use is not permitted by statutory regulation or exceeds the permitted use, you will need to obtain permission directly from the copyright holder.

Informal Reuse, Repair and Refurbishment Business Networks for Air Conditioners in Gangxia Village, Shenzhen

Yuk Tung Chow and Benjamin Steuer

Abstract Circular Economy (CE) related approaches have emerged as a central strategy among urban governments and corporate actors involved in waste management (WM) and have been a focus of the China's administration [23]. For both domains the informal recycling sector (IRS) (By 'informal' the chapter refers to any stakeholder that does hold some but not old all, officially required registrations to deal with WEEE) has been pivotal: Be that for shaping value chains of urban China's WM or in achieving CE-styled life-time extensions for more complex consumer goods. The parallelism of the formal, state orchestrated and the informal systems have led to a critical challenge for circularity in China. Both sides operate within their respective systemic (rule-based) frameworks, and the informally developed system around second hand electronics effectively serves demands of the floating population in urban villages, such as Gangxia Shenzhen. With growing urbanization and official formalization contesting dynamics have emerged between municipal administrators and the IRS. This raises the question of why local policies did not organically integrate this recycling sector via the official top-down approach. The inquiry pursued in this chapter shows how informal recyclers do in some instances contribute to a more inclusive urban WM system that features social-environmentally sustainable CE practices beyond recycling. Moreover, the complex social fabric of this sector is inseparably intertwined with aspects of urbanization and employment options that lie at the root of the city's complex development pattern.

Keywords Circular economy · E-waste · Repair · Reuse · Informal economy · China

---

Y. T. Chow · B. Steuer (✉)
Division of Environment and Sustainability, The Hong Kong University of Science and Technology, Kowloon, Hong Kong
e-mail: bst@ust.hk

Y. T. Chow
e-mail: yuktungchow@gmail.com

© The Author(s) 2026
M. de Jong et al. (eds.), *The Inclusive Circular Economy*, Urban Sustainability,
https://doi.org/10.1007/978-981-96-6867-0_10

1 Introduction

This chapter aims to provide an insight from a case study in the urban village of Gangxia in Shenzhen. Within this urban ecosystem, a small-scaled, autonomously developed recovery sector for air conditioners (ACs) symbiotically coexists with formal enterprises that form a circular loop, which provides social and material benefits to local residents. The study finds that this sector meets the floating population's demand in Gangxia for second-hand ACs by applying CE strategies that avoid deformative recycling and, by implication, material and value losses. Informal recycling sector (IRS) actors change their operative status in line with policy developments and have in some cases abandoned fully informal practices in favour of more formalized structures. Some stakeholders established small and medium-sized enterprises (SMEs), while others maintain part time jobs along the value chain.

Roughly, the present chapter pursues three research questions: First, what intersections between the formal and informal system exist in Gangxia's CE on ACs? Second, which institutionalised patterns does the informal segment employ when operating their circular business models for AC reuse, repair and refurbishment? Third, to which extent have local policies integrated or provided leeway for informal practices in this context? In line with these questions, the chapter provides insight into CE activities around ACs in Shenzhen's Gangxia village. In detail, we aim to illustrate how the existence of formal and informal rule systems shape these activities. As a result we intend to show how policies on appliance reuse, repair and refurbishment have shaped actual practices and IRS stakeholders' identities in Gangxia village.

The paper proceeds by outlining the methodology for the case study in Shenzhen. The first empirical section provides an overview on WEEE generation, related policies and the IRS in China. We then proceed by analysing the policy environment for WEEE management at national and local levels before diving into the discussion part. Here we first highlight the impact of formal institutions on the informal segment's operations before analysing the institutional foundations and the operational practices of informal CE business models on second-hand ACs.

2 Materials and Methods

2.1 *Means of Data Gathering*

With the aim of analysing circular business model within small-scaled loops, field work was conducted in June 2021 in the oldest urban village Gangxia in Shenzhen. Through field research that encompassed over 50 hours of observational evidence gathering, a multitude of short individual communications and several, semi-structured interviews with circular business managers were conducted to understand how policy had shaped informal business activities. In detail, 10 semi-structured interviews were conducted: Seven with self-employed individuals (getihu 个体户),

Table 1 Survey questions asked to respondents in the WEEE refurbishing and reuse sector in Gangxia village, Shenzhen

Interview topics	Survey questions
Background information	Scale of employment, number of people, area and population covered by recycling channels
Reason and factors behind the change in formal regulations	Do policies, urban management and other government organizations, the intervention of large platforms, and market competition impact urban transformations?
	What about limitations of others (from your family, village, personal environment)? Did you engage in what you do because you were inspired by others?
Specific policy influence on the business	In 2011, it was promulgated that the disposal of waste electrical and electronic products should be recycled and dismantled through formal channels. Does it affect the source of business channels and costs? Does it also affect the way you engage in your business? If yes, what did you change in your business activities?
Recycling history	Did you have relevant experience before establishing a formal recycling company? If yes, what are the changes in business models and recycling costs; if not, why did you join the recycling industry?
	When did you start to rent a house in the urban villages to start/continue collecting WEEE?
	Application of tools: Did you use new tools or increase the use of manpower in the process of your business transformation?

who operate AC refurbishment and reuse shops in Gangxia village and three with furniture movers, who help assembling ACs. The question items inquired among the interviewees are listed in Table 1. All information and data for this chapter were gathered during field work from May to June 2021.

2.2 Contextualisation of the Survey Area

Regarding the context of the researched area, Shenzhen, as a young city with a highly mobile population, constitutes a highly unique case. According to the recent census [32], 80% of the population are migrant labourers without a registered permanent residence (户口 *hukou*). The urban villages in the city (*chengzhongcun* 城中村) are often their first place to stay, which is where the demand for second-hand electrical appliances mainly emerges. In Futian District, non-registered permanent residence accounts for 59.78% of the whole population, according to the Futian government website [10]. Since the Reform and Opening period, the collaborative business model based on the kinship has been developed in the electronics refurbishing business segment [3].

The newly migrated residents have a common demand for second-hand home appliances such as ACs and washing machines. Second-hand household appliances in Gangxia village are often repeatedly traded until they have reached their end-of-life. In an effort to counter this dynamic, the internal recycling chain of household appliances in Gangxia village practices multiple cycles of ownership transfer. Herein, repaired devices are exchanged between initial users, local repair and trading shops and subsequent users before being ultimately discarded. When a new user enters the Gangxia community and requires refurbished, second-hand home appliances, he/she can obtain information on provisioning sources from other residents in the community. Upon acquisition of the device at a second-hand home appliance trading point in the community products will be provided via a delivery service. In the case of Gangxia, official policies fall short of registering informal practitioners which are therefore not evaluated in regard to their contribution to the local CE. Overall, informal practitioners vastly contribute to reuse practices, and dense practitioner clusters benefit cooperative networks and develop the social exchange structures.

3 Overview on WEEE Generation, Policies and the Informal Recycling Sector in China

Solid waste is a major global challenge that particularly grows in substance during the process of urbanization [7, 14, 15]. Within the municipal solid waste (MSW) stream, waste electrical and electronic equipment (WEEE) has emerged as an increasingly challenging stream. In China, WEEE generation has exhibited an extremely strong growth rate since the mid-1990s to the mid-2010s. Estimations by the China Household Electronic Appliances Research Institute (CHEARI) indicate that this growth trend is led by mobile phones (from approx. 200 to 1,200 million units over 2003–2017), television sets (from approx. 200 to 600 million units over 1998–2017) and ACs (from approx. 0.2 to 400 million units over 1996–2017) [44]. In terms of recovery, officially documented quantities have recorded respectively high numbers in absolute terms, while growth rates have in turn exhibited a marginal decline, particularly in 2022 (Fig. 1). While various explanations potentially help to explain this trend, the pivotal aspect is probably the saturation of the Chinese household electronics consumer market.

Informal recycling activities covering collection, dismantling, material extraction and various recovery and reuse practices of MSW and WEEE, are central factors in urban China's waste management [13, 17, 38, 39, 43]. Many empirical studies have outlined the IRS' contribution to mitigating the waste challenge [1, 11, 13, 41]. However, current policy in China does pay little attention to the advantages of informal CE practices that go beyond recycling [38]. This policy shortcoming is to some extent a global phenomenon as even more environmentally focussed, international regulatory frameworks, such as the EU's, have began to develop guidelines for

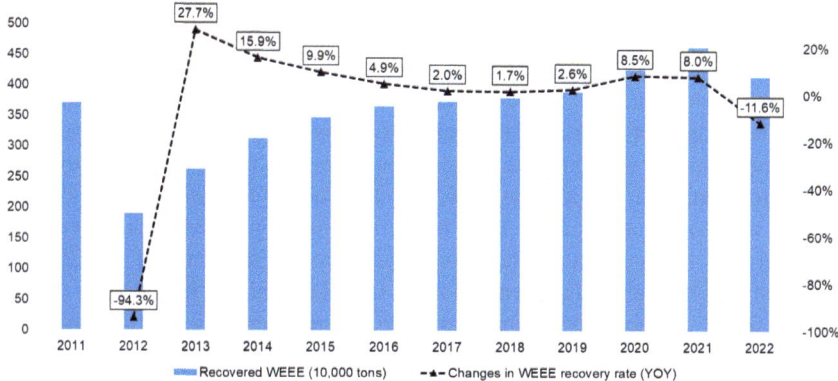

Fig. 1 Officially documented quantities of recovered WEEE in China (based on [18])

electronic product reuse and repair [6]. A major driver for such alternatives to recycling and, in the sense of the CE more beneficial approaches, is the stronger impact on safeguarding material values and the extension of the monetary value chain. Based on observations in the Netherlands [28], it has been pointed out that smaller circular loops advance the CE [35]. Herein, practices that lead to reuse rank notably high in the sense of maintaining the value and utility of material and product stocks [35]. In comparison to recycling, reuse furthermore reduces environmental externalities and helps extending the service life of electronic appliances [22]. In this context, CE SMEs play a particularly important role. However, due to their weak capital endowment, reuse centred SMEs require government support for maintaining operations [45]. The relatively flexible, community-based WEEE-repair-resell service models in China that fall in this category can effectively meet the needs of users by shifting from resource-intense and wasteful to employment-intense business models [35]. However, in China, as is the case for the international realm, SMEs in this field have not been adequately addressed by relevant policy measures [27].

So while the CE serves as the main strategy of China's policy on WM, aiming to maximize the use of resources via enlarging the potential economic value derived from the recycling industry [27], many active stakeholders such as repair-centred SMEs have not benefitted from official resource allocations. Moreover, the management regulations in various Chinese cities do not accommodate the presence of this group [1, 39]. Part and parcel of this negligence is that such, often IRS affiliated, business models conflict with the management of city appearances, leading to sometimes conflicting development patterns of interaction [39]. However, some recent policy approaches tend to open up to the informal segment leading to an effective integration of the IRS in secondary material management [41]. Such approaches are subject to the larger process of urbanization and how segments are fiscally or policy-wise supported, which is needed to prevent valuable patterns in the sense of the CE to come under the wheels of competition and disappear.

A better understanding about local practices and capabilities of repair and refurbishment SMEs entails various benefits—not only for research, but moreover for assessing resource management and economic dimensions of the sector as well as for policy implementation. Since formal recycling plants largely source materials from informal collection, tracing the sources for this waste stream will contribute to the estimation of recycling efficiency [14]. Furthermore, it is necessary to evaluate the environmental benefits of reuse [16, 29]. As reuse is central to maximizing the residual value of resources at the local level, forming a small-scale material trading cycle between regions is a key task for urbanization in the sense of the CE [16]. Many scholars have also suggested that if the size of circular exchange loops were to be reduced, the respective economic and environmental benefit would increase [1, 2, 8, 25]. In Shenzhen's Gangxia village, we find ample evidence on circular business models around electronic appliance reuse, repair and refurbishment, which serves local demand for second-hand use.

4 The Policy Environment for WEEE Management in China and Shenzhen: Recycling vs. Repair, Refurbishment & Reuse

4.1 National WEEE Regulations

In regard to environmental protection and, more specifically WM, regulatory frameworks remained relatively salient up until the mid-1990s. Since then, research has documented a full-fledged wave of institutionalisation (rule-creation), which continues until the present day. As a result, China's institutional environment for WM is characterized by a plethora of laws, directives and regulations, with often weak levels of effectiveness and in some cases notable loopholes [38, 39].

The regulatory scaffolding for specific waste streams such as WEEE features two notable particularities pertaining to (1) legal sources and (2) institutional (rule-systemic) refinement. In regard to the former, China's WEEE regulations are sourced from different origins. First, China has essentially copied foreign regulations such as the European Union's Regulation on Hazardous Substances (RoHS) and it's WEEE directive via a locally-adjusted institutionalisation process[1] [30]. Second, to account for domestic characteristics, China achieved significant advancements in legislation through local pilot policy experiments aiming to refine national legislation and fine-tune WEEE management practices. This dynamic was especially prominent between 2003 and 2011 and lead to new recovery mechanisms such as the Old-for-New WEEE trade-in scheme or China's WEEE fund to sustain downstream processing [31].

---

[1] In China, the respective regulations are *'The Regulation for the Control of Pollution caused by Electronic Information Products'* and *'The Regulation on Management of Waste Electrical and Electronic Equipment, Recycling and Disposal'*.

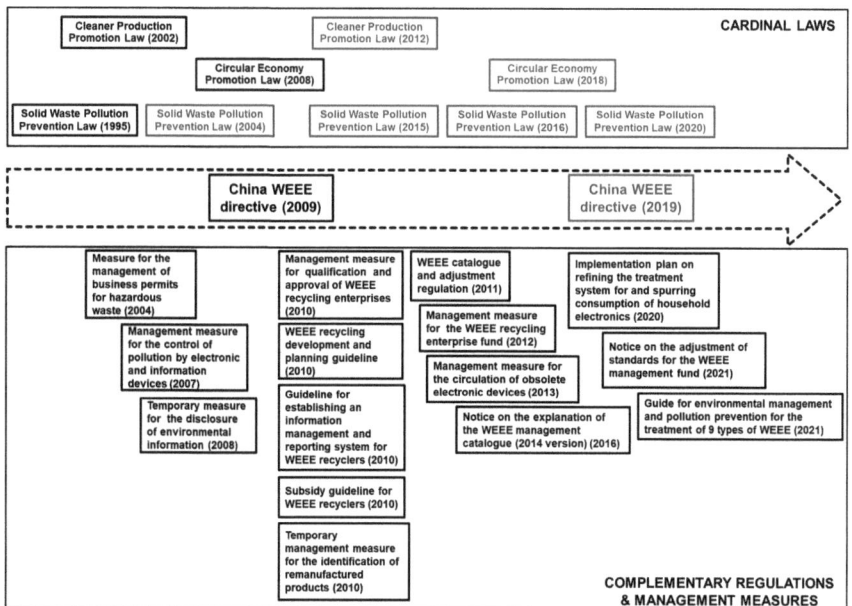

Fig. 2 Institutional evolution of China's WEEE management system (issuance dates in grey indicate amended versions)

These efforts culminated in a formally comprehensive, rule systemic refinement for WEEE governance. Herein, China's WEEE directive constitutes the central element of the regulatory system, which is (1) inspired by governance strategies stipulated in superordinated, cardinal laws and (2) further refined by subordinated, complementing measures (see Fig. 2).

Within this framework, the CE plays a key role. Adopted by the Chinese government to increase resource efficiency and reduce anthropogenic environmental impacts, the CE was first trialled in pilots during the early 1990s and formally institutionalized as Circular Economy Promotion Law in 2008. This key law sets forth central aims of China's WEEE governance that influence management practices and structure until the very present. Discarded electronic devices are to be handled via recycling and reprocessing measures, whereas reuse and refurbishment approaches only receive minor attention [23, art. 39]. This treatment strategy reduces CE ideals to mere recycling, which is a prevalent trend among most international CE systems. It, however, critically fails to recognise the gains from more material and value conserving CE approaches, such as remanufacturing, refurbishment, repair and reuse patterns [28]. From a causality-centred, systemic perspective this policy ideal entails substantial challenges. Adhering to end-of-pipe approaches that focus on managing a given problem, switching to alternative CE strategies would help to prevent or mitigate the emergence of a problem in the first place.

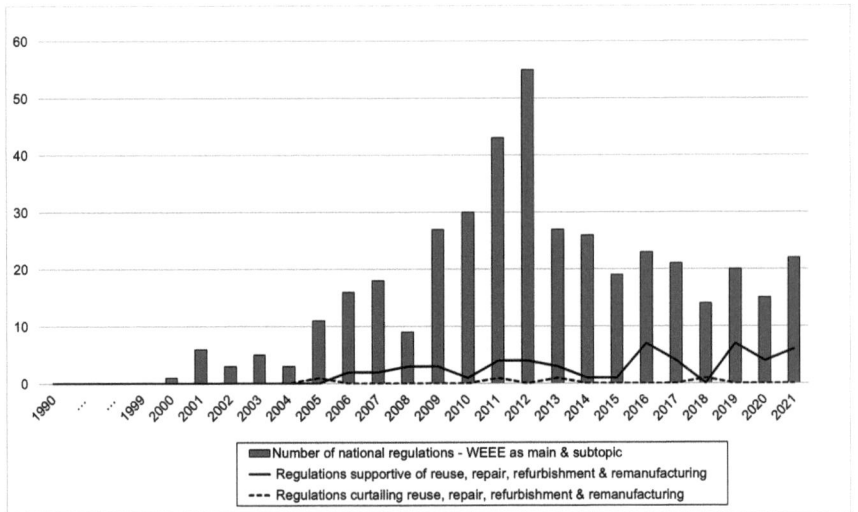

Fig. 3 National regulations on WEEE management with particular focus on reuse, repair, refurbishment and remanufacturing (the author based on www.lawinfochina.com)

Against this background, analysing China's recent trend of institutionalisation and management reveals an interesting shift. While initial pilot efforts between 2003 and 2012 have mostly focussed on identifying recycling-centred approaches, yet still infant trends indicate the rise of alternative CE practices. From a purely quantitative perspective, it can be discerned that rule-making for WEEE has been less vibrant after the year 2012, in which the WEEE recycling fund to support formal processing companies was issued. In close temporal proximity, however, regulations that support reuse, repair, refurbishment and remanufacturing practices have grown in prominence (Fig. 3). This begs the question of a potential change in mindsets among policy-makers in favour of more material and economic value conserving strategies.

First indications of this trend emerged in the WEEE directive (2011), stipulating the legality of repair and reuse of old electronic devices given proper labelling [36, art. 12 and art. 49]. Falling short of more substantial steps in this direction, the recently amended version of the directive is mostly focussed on updating terminologies, instead of new outlining new management approaches [37]. More specific trends towards legally supporting repair and reuse practices are notable in the Management Measure on the Circulation of Obsolete Electronic Devices. Herein, individual entrepreneurs, often belonging to the informal or non-registered sector, are acknowledged as traders of refurbished WEEE [18]. Additionally, processes and standards pertaining to repair, product labelling and transaction documentation are stipulated (ibidem, art. 3, 7, 8, 13, 15, 17). Further sustaining these CE strategies is the 2014 Notice on the value-added tax (VAT) Collection policy that reduced the VAT for selling second hand electronic devices from 4 to 2% [5].

Of more recent importance, that is since 2016, two major institutional innovations have shaped China's WEEE management in regard to reuse and repair. The first aspect pertains to the significant reductions in subsidies for formal WEEE processors. Initially in 2012 relatively high, appliance unit based subsidies were provided via the China WEEE fund to develop modern processing capacities among selected recycling facilities. These were adjusted in 2016, potentially in line with experiences collected during the first four years of the scheme, before substantial subsidy reductions occurred in 2021 (Fig. 4). This latter measure constitutes a *caesura* to the comparatively generously allocated official support. One assumption is that this change can be traced back to official disapproval as subsidies were mostly used by processors to buyback WEEE from informal collectors [31]. A direct implication of this adjustment is that recycling has become less profitable for formal companies. Consequently, alternative CE approaches such as repair and refurbishment now promise higher economic turnover to operators, which potentially entails bad news for recycling facility. Take for example a collector, who could sell devices either to recyclers or to repair, refurbishing and remanufacturing entities, which are mostly belonging to the informal domain. Given the 2021 enacted reduction in unit subsidies, i.e. the base value of what recyclers could pay to collectors, collectors with an eye on profits, might be more prone to sell it to (informal) repair shops than to formal recyclers.

Second to that is the 2020 Plan to revitalize recycling and consumption of household appliances. Issued by the powerful National Development and Reform Commission, the scheme features entirely novel approaches towards WEEE that

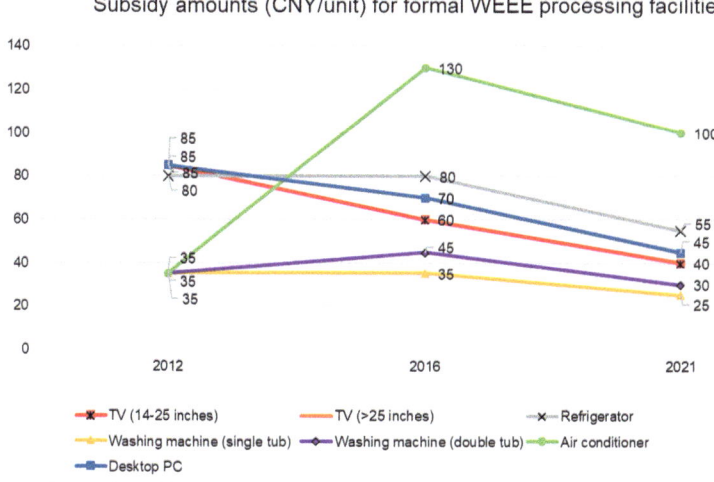

Fig. 4 Subsidies (CNY/unit) for formal WEEE processing companies (based on MOF et al. [19–21])

envisage CE principles beyond recycling. First, repair services by individual (non-registered, informal) entities at the household level has for the first time become an acknowledged and promoted strategy. Second, the document suggests formal recycling companies 'recruit' (*xishou* 吸收) these stakeholders and integrate them into their business operation [24, art. 1 and 2]. Third, the document promotes the development of second hand markets for refurbished and remanufactured electronics (ibid., art. 5), which have traditionally been dominated by non-registered, informal traders [38]. Forth, the commercialisation of stewardship based business models for electronic devices (*jiadian zulin xin xingye* 家电租赁新兴业) and the development of respectively supporting frameworks are substantially highlighted (ibid., art. 6). The caveat, for the by and large informally dominated sector, is the equally strong emphasis on registration and documentation (ibid., art. 9). In sum, both recent policy measures appear to indicate a shift away from a sole support of WEEE recycling, and towards a support of alternative CE practices (refurbishment, repair and reuse) whilst including grey or informal stakeholders.

4.2 WEEE Regulations in Shenzhen

One of the major striking aspects of WEEE legislation in Shenzhen is the city's notable institutional inactivity. In contrast to key WEEE pilot areas that extensively promoted WEEE regulations during 2003–2011 [31]), Shenzhen has been relatively quiet on the subject matter. While the municipality has followed suit to the national WEEE directive in 2011 and adopted the Old-for-New Household appliance trade in scheme more recent policy measures on the topic are relatively sparse: Nothing has been issued on CE approaches such as reuse, refurbishment or repair of WEEE. Yet, there are some indications that the local government envisages a more inclusive role for the IRS in managing WEEE: First, Shenzhen's 2020 Household Waste Separation Guideline—a regulatory measure issued by all major municipalities in the wake of Xi Jinping's increased emphasis on waste management [40]—stipulates that IRS stakeholders are allowed to act as collectors for WEEE discarded by households [34, art. 17]. The same document also states that all downstream processing entities need to be registered and have to hold proper official qualifications (ibid., art. 31), which effectively excludes the IRS from recycling, repair or refurbishing activities. Second, Shenzhen's Bureau of Commerce published a public consultation document on a WEEE recovery management system in 2021. Here again the IRS is considered as a viable operator for WEEE collection [32, art. 2] and multi-channel collection—an explicit allusion to informal and formal collection structures [31]—is explicitly encouraged (ibid., art. 5). However, similar to earlier policies, collecting stakeholders are not allowed to further conduct any deformative WEEE processing activities (ibid., art. 10).

Overall, the national and local regulatory setting entail some implications for the IRS in Shenzhen: While the sector is clearly considered as viable operator for collection, policies clearly try to keep processing activities in line with a strictly

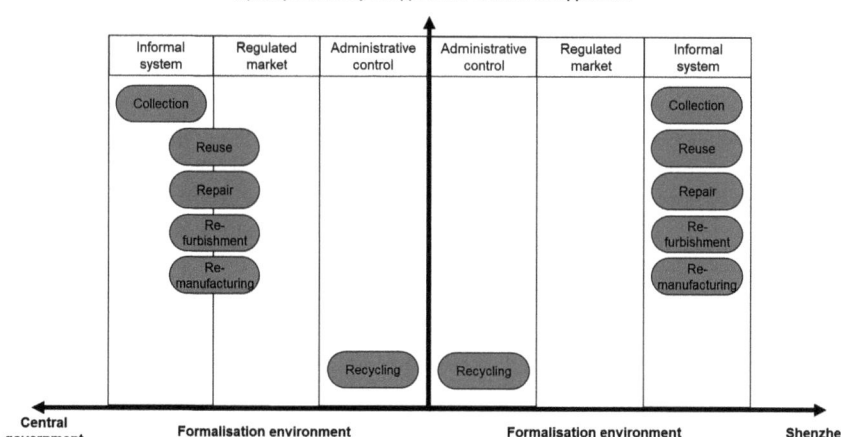

Fig. 5 Central government and Shenzhen's formalisation of WEEE management along the R-principles

monitored and controlled formal system. What therefore remains relatively unregulated are those CE practices such as reuse, repair, refurbishment and remanufacturing that hierarchically precede recycling. In this regard, national regulation seems to have become increasingly accustoming towards actual practices of the IRS, while Shenzhen's regulatory strategy seems to essentially restrain itself and allow market dynamics to follow demand. By implication, the governmental actors at central and local levels have rather focussed on institutionally controlling recycling, which ranks relatively low in along the CE's R-principle hierarchy [28], whereas higher ranking R-practices are being left subject to market dynamics (Fig. 5).

5 Discussion of the Formal-Informal Dynamic in Gangxia

5.1 The Business Model of Gangxia's Informal Sector in AC Refurbishment and Reuse

Although not all players in the domain of AC reuse are fully integrated into the value chain, the essential circular business model in Gangxia centres on serving the short-time market demand of migrant labour. Most of these stakeholders live in rental houses in urban villages and have a short-period demand for affordable home appliances such as ACs. Here, the short-term demand for electrical appliances has promoted the rise of reuse, repair and refurbishing businesses that focus on such specific market demand.

It is notable that the main business concept in Gangxia is direct reuse buttressed by occasional repair and refurbishment. Due to low operating volumes and running costs, operating actors have access to, take-back and resell a relative constant amount of second-hand home appliances with a high circulation frequency, which leads to a use-intense appliance reuse network in Gangxia. Consumers, largely represented by low-income migrant population, benefit from access to low-cost second-hand appliances and are thereby effectively served by local reuse, repair and refurbishment models. To operate product circulation and transaction of ACs and other second-hand home appliances, local operators have developed a division of labour within a relatively narrow, geographical: After collection, devices are sent to repair shops before being resold or in case of substantial malfunction dismantled. Single operative instances exhibit a certain degree in division of labour (Table 2).

In regard to what could remotely be described as internal transaction and labour force management, Gangxia's business model exhibits several notable characteristics. First, its collaborative business network is foremost based on geographical relationships and gradually formed exchange patterns within. This comprises two stakeholder groups, which mostly act individually as self-employed circular operatives. Among them, collectors as first group are mostly composed of part-time active, migrant workers from Sichuan, who gather materials at the entrance of the village. Their main occupation consists of providing residents in the urban villages with moving services for household furniture and other goods within a short distance. If in the process appliances are found to be in the need for repair, collectors would transfer these to self-employed entrepreneurs (*getihu* 个体户). This second group of self-employed circular operators rely on little capital input for cleaning and repair the appliances before selling these as second-hand products to other users in urban villages. In addition, a third yet sporadically occurring group of itinerant recyclers appear in urban villages such as Gangxia. They will move in-between different areas,

Table 2 Division of labour along the product value-chain for AC repair, refurbishment and reselling practices by informal entrepreneurs

Role	Description	Labour size	Equipment
Delivery & collection	Home movers deliver ACs to Gangxia, to designated collectors; Movers in charge of installation and disassembly of ACs; part-time job;	Up to 8 persons	Two shared trucks and tollers
Repair & refurbishment	Supplied by in-house collectors, second-hand shops repair and clean ACs; Division of labour centred on kinship/ family. Storage of ACs after refurbishment	Approximately three persons per shop; additionally, two part-time family members for each shop	Rented shop
Reselling	Various shops, clustered within geographically designated location	Varying	Rented shop

Table 3 The operation costs of AC repair and resell shops

Fixed cost	Amount (CNY/month)
Rent (median value)	5000
Utility	200
Transportation (within the urban village)	300
Total	5500

mainly aiming to acquire discarded electronics for sale to material processors outside of the city for material extraction.

Second, the circular businesses around ACs and other electronics in Gangxia feature a strong element of kinship and family based working relations. Historically, this is rooted in initial informal economy structures that emerged in Shenzhen alongside the capital-intensive urbanization process since the 1980s. Arriving migrant workers launched collective-business patterns based on social trust [42], and gradually became the dominant actors valorising recyclable waste streams that continuously grew in line with urban expansion [3, 12]. In the process of forming material waste exchange relationships, networking between fellow villagers turned out to be a central element in cooperative partnership development in Gangxia's community. Within each sector the repair and reuse chain, individual groups of people emerged that share the same hometown or are bound by kinships. According to interviews with involved stakeholders, kinship-based partnerships increasingly permeated the entire segment resulting in a sometimes hierarchically arranged division of labour within the repair and refurbishment business network (as shown in Table 2).

Third, familiarity and kinship has had an effect on trust-building and in turn capital good sharing to advance common business objectives. For example, conducting long-distance transports would be done by renting one member's individual truck; sudden labour shortcomings would be tackled by interim replacement by another member's relatives. Such flexible and resource pooling helps reducing transportation and other fixed expenses and thus improves operational efficiency, particularly in regard to material use and transaction costs (e.g. Table 3). In the same manner as resource inputs and labour are shared, family members will receive incomes in alignment with their work contributions. Given that workloads are relatively minor for typical ACs taking around 2–3 h, daily repair volumes can be easily adjusted based on available labour and time capacities.

5.2 The Influence of Formal Policy on Informal CE Business Operations

The policy environment has played a formative role in shaping operations of circular businesses in the urban village of Gangxia. Generally two trends in urban governance of the IRS could be identified. First, the municipal government perceives stakeholders engaged in circular management of AVs as jeopardising public safety and as an obstacle to the city's cultured appearance.

In regard to the former, the *Decision of the Standing Committee of the Municipal People's Congress on Resolute Investigation and Punishment of Illegal Buildings* (1991–2004) and the *Opinions on the Handling of Historical Illegal Private Houses in the Shenzhen Special Economic Zone* (2002), aimed at modernising the urban appearance of the village. In turn, older buildings inhabited by labour migrants and members of the IRS were torn down and induced a reduction in living space for these stakeholders. As a consequence, some IRS stakeholders moved out of the area to find a more accommodating business environment elsewhere. Others left as they could not afford the high shop rents. As an exception only those with sufficiently strong capital backing remained, set up shops, and in line with policy requirements, registered as self-employed entrepreneurs (*getihu*, 个体户). As for those practitioners, who relied on stalls and booths for their repair business, the *Shenzhen Special Economic Zone City Appearance and Environmental Sanitation Management Regulation* [33] entailed a curtailing effect on operations: The regulation prohibited any unorderly storing or stacking of tools or devices, which limited operational space and in turn drove out itinerant stakeholders engaging in appliance repair and refurbishment.

In general, local regulations my not directly target, but rather indirectly impede IRS activities. For collectors recovering and transporting used electronics, obtaining official permits as individual entrepreneurs (*getihu* 个体户) is nearly impossible. The reason is that these need to possess a rental agreement for a working place to obtain this type of business registration. Similarly, itinerant furniture movers can not formally register their business and are therefore excluded from policy conceptualisation. This lacking recognition of their services by the government, negatively effects a key segment of the IRS in Gangxia's appliance repair and refurbishment. Secondly, current subsidy arrangements for circular management of electronics is only accessible to large-scale, accredited WEEE recycling facilities. As the repair-for-reuse domain is exempted from these financial provisions, broadening the CE and upscaling practices beyond recycling for (near) EOL devices is impeded. Additionally, price guidelines for repair and refurbishment services of second-hand electronics were set relatively low being perceived as curtailing potential profit margins of IRS operators (personal communication, repair show owner, Gangxia, 25 June 2021). More recent subsidy provisions at the local level, which follow the national trend on promoting the formalisation of WEEE refurbishment and remanufacturing equally aim at driving out the informal repair shops [4]. In final instance, a regulatory heritage of the Maoist period, the IRS has opted out of the permanent residence registration system or hukou, which effectively binds an individuals' access to welfare services to their registered hometown. Informal stakeholders, labour migrants by virtue, can not benefit from any social security services, which in turn further complicates their operations in the CE if done outside of their registered locus of residence.

6 Conclusion

The case study of informal CE operators for AC repair and refurbishment in Shenzhen's Gangxia village demonstrates the network strength, CE capacity and service value for other low-income groups. By design, it can prevent the premature end-of-life of second-hand household appliances, effectively reduce the ecological footprint, and generate social and environmental benefits. A main driver behind the network's continued existence is the vast demand for short-time use of second hand household appliances from labour migrants.

While national and local policy environments largely neglect higher ranking CE practices, e.g. reuse, repair, refurbishment and remanufacturing, the local IRS has made these practices into their main circular business model for managing ACs. To By virtue of a set of informal rule systems, the IRS has developed a relatively effective business concept. These essentially comprise (1) kinship and familiarity, to enhance internal reliability to facilitate transactions; (2) service provisions to enable the availability of products and thus serve the needs of users; (3) adaptive resilience and flexibility to formal policy measures;

With regard to these three aspects, the internal recycling network of second-hand ACs has its unique advantages. Inspired by Peluso and Ribot [26]'s paper about mapping the access to natural resources within the local context, our case study reflects the significance of local social relationships. Such connections are embodied in the organization of behavioural patterns that result in monetarized material transactions centred on second-hand household appliances. Given that current policies sideline actors outside the formal realm, more empirical studies are needed to identify how inter-human relationships respond to formal regulatory requirements. An understanding of these micro-dynamics is needed to render models and numerical data into meaningful assessments of real time CE systems. Since SMEs are essential to promote green jobs and economic transitions in the supply chains [9], tailored regulation instruments would encourage a more internalized and inclusive institutional surrounding for the local participators. Therefore, it is recommended that concepts like the zero-waste city policy as part of the CE strategy should strive to do more towards resolving the exclusive nature of current policy settings in Gangxia. Doing so would entail substantial benefits: The informal segment not only realises CE structures that go beyond the overemphasised formal economic preference for recycling and thereby contribute to conserving resource and product values. Moreover, its mere existence creates benefits for the social and economic dimensions of sustainability such as employment opportunities and product supply. To buttress these advantages the official system could provide training, official certification and product safety management assistance, which in turn constitute a step towards socio-economic integration and sustainable city development. On the long term such measures that ideally enable formalisation on terms of informal stakeholders hold potential to advance Shenzhen one further along a trajectory towards an inclusive, waste-wise city.

References

1. Barford A, Ahmad SR (2021) A call for a socially restorative circular economy: waste pickers in the recycled plastics supply chain. Circ Econ Sustain. https://doi.org/10.1007/s43615-021-00056-7
2. Calisto Friant M, Vermeulen WJV, Salomone R (2020) A typology of circular economy discourses: Navigating the diverse visions of a contested paradigm. Resour Conserv Recycl 161:104917. https://doi.org/10.1016/j.resconrec.2020.104917
3. Chen Y-C, Chen M-N (2020) Social trust and open innovation in an informal economy: the emergence of Shenzhen mobile phone industry. Sustainability 12(3):775. https://doi.org/10.3390/su12030775
4. Chinadaily.com.cn (2022, May 16) Shenzhen encourages rule-conform mobile phone refurbishment Recycling Love and other second-hand platforms welcome the policy (深圳鼓励合规翻新手机销售 爱回收等二手 3C 平台迎政策利好) (Online). https://caijing.chinadaily.com.cn/a/202205/26/WS628f2510a3101c3ee7ad7563.html. Accessed 10 Sept 2023
5. China Ministry of Commerce (MOC) and China State Bureau of Taxation (CSBT) (2012) Notice on the Streamlining and Combination of Value Added Tax Levy Rates (in Chinese). http://www.chinatax.gov.cn/n810341/n810765/n812141/n812252/c1078631/content.html. Accessed 10 Apr. 2016.
6. Circular Economy Action Plan—For a cleaner and more competitive Europe (2020) European Commission (EC). https://ec.europa.eu/environment/strategy/circular-economy-action-plan_de
7. Cooper DR, Gutowski TG (2017) The environmental impacts of reuse: a review. J Ind Ecol 21(1):38–56. https://doi.org/10.1111/jiec.12388
8. De Pascale, A., Arbolino, R., Szopik-Depczyńska, K., Limosani, M., & Ioppolo, G. (2021). A systematic review for measuring circular economy: The 61 indicators
9. Decent Work, Green Jobs and the Sustainable Economy (2015) http://www.ilo.org/global/publications/books/WCMS_373209/lang--en/index.htm
10. Futian Government Online (2012, April 10) Research on the relationship between non-registered population and Futian's economic and social development. http://www.szft.gov.cn/bmxx/qtjj/zwxxgk/tjxx/jjshyxts/content/post_4202577.html
11. Gall M, Wiener M, Chagas de Oliveira C, Lang RW, Hansen EG (2020) Building a circular plastics economy with informal waste pickers: recyclate quality, business model, and societal impacts. Resour Conserv Recycl 156:104685. https://doi.org/10.1016/j.resconrec.2020.104685
12. Goldstein J (2017) Just how "wicked" is Beijing's waste problem? A response to "the rise and fall of a 'waste city' in the construction of an 'urban circular economic system': The changing landscape of waste in Beijing" by Xin Tong and Dongyan Tao. Resour Conserv Recycl 117:177–182. https://doi.org/10.1016/j.resconrec.2016.10.018
13. Goldstein J (2020) Remains of the everyday: a century of recycling in Beijing. University of California Press
14. Huang H, Tong X, Cai Y, Tian H (2020) Gap between discarding and recycling: estimate lifespan of electronic products by survey in formal recycling plants in China. Resour Conserv Recycl 156:104700
15. Kirchherr J, Reike D, Hekkert M (2017) Conceptualizing the circular economy: an analysis of 114 definitions. Resour Conserv Recycl 127:221–232. https://doi.org/10.1016/j.resconrec.2017.09.005
16. Kristensen HS, Mosgaard MA (2020) A review of micro level indicators for a circular economy—moving away from the three dimensions of sustainability? J Clean Prod 243:118531. https://doi.org/10.1016/j.jclepro.2019.118531
17. Li S (2002) Junk-buyers as the linkage between waste sources and redemption depots in urban China: the case of Wuhan. Resour Conserv Recycl 36(4):319–335. https://doi.org/10.1016/S0921-3449(02)00054-X

18. Ministry of Commerce (MOC) (2012-2023) Annual report on China's renewable resources recovery industry (in Chinese). Online: http://www.crra.com.cn/searchResult?query=中国再生资源回收行业发展报告
19. Ministry of Finance (MOF) et al. (2012). Management measure on using the WEEE fund levy (Version 2012) (34/2012)
20. Ministry of Finance (MOF) et al (2015) WEEE management fund subsidy standard (91/2015)
21. Ministry of Finance (MOF) et al (2021) WEEE management fund subsidy standard (10/2021). Ministry of Finance (MOF) et al.
22. Moraga G, Huysveld S, Mathieux F, Blengini GA, Alaerts L, Van Acker K, de Meester S, Dewulf J (2019) Circular economy indicators: What do they measure? Resour Conserv Recycl 146:452–461. https://doi.org/10.1016/j.resconrec.2019.03.045
23. National People's Congress (NPC) (2008) Notice of the state council on printing and distributing the national environmental protection "eleventh five-year plan."
24. National Reform and Development Commission (NDRC) (2021) Notice on the implementation plan for improving the recovery system for discarded household devices and on promoting the replacement and consumption of household electronics (752/2020)
25. Padilla-Rivera A, Russo-Garrido S, Merveille N (2020) Addressing the Social aspects of a circular economy: a systematic literature review. Sustainability 12(19):7912. https://doi.org/10.3390/su12197912
26. Peluso NL, Ribot J (2020) Postscript: a theory of access revisited. Soc Nat Resour 33(2):300–306. https://doi.org/10.1080/08941920.2019.1709929
27. Pesce M, Tamai I, Guo D, Critto A, Brombal D, Wang X, Cheng H, Marcomini A (2020) Circular economy in China: translating principles into practice. Sustainability 12(3):832
28. Potting, J., Hekkert, M., Worrell, E., & Hanemaaijer, A. (2017). *Circular economy: Measuring innovation in the product chain*. PBL Publishers.
29. Saidani M, Yannou B, Leroy Y, Cluzel F, Kendall A (2019) A taxonomy of circular economy indicators. J Clean Prod 207:542–559. https://doi.org/10.1016/j.jclepro.2018.10.014
30. Salhofer S, Steuer B, Ramusch R, Beigl P (2016) WEEE management in Europe and China—a comparison. Waste Manage 57:27–35
31. Schulz Y, Steuer B (2017) Dealing with discarded e-devices. In: Routledge handbook of environmental policy in China. Routledge, pp 314–328
32. Shenzhen Bureau of Commerce (SZBC) (2021) Shenzhen WEEE recovery system (Consultation Document). http://commerce.sz.gov.cn/hdjlpt/yjzj/answer/16268
33. Shenzhen Municipal Bureau of Ecological Environment (SMBEE)) (2019) Shenzhen special economic zone city appearance and environmental sanitation management regulations—policies and regulations. http://www.sz.gov.cn/cn/xxgk/zfxxgj/zcfg/content/post_8968657.html
34. Shenzhen People's Congress (SZPC) (2020) Household waste separation management ordinance. http://commerce.sz.gov.cn/hdjlpt/yjzj/answer/16268
35. Stahel WR (2019) The circular economy: a user's guide (1st ed.). Routledge. https://doi.org/10.4324/9780429259203
36. State of Council (SC) (2011) Opinions of the general office of the state council on establishing a complete and advanced recycling system for used commodities
37. State of Council (SC) (2019) Notice of the general office of the state council on printing and distributing the pilot work plan for the construction of "waste-free cities."
38. Steuer B (2016) What institutional dynamics guide waste electrical and electronic equipment refurbishment and reuse in urban China? Recycling 1(2):286–310
39. Steuer B (2017) Is China's regulatory system on urban household waste collection effective? An evidence-based analysis on the evolution of formal rules and contravening informal practices. Journal of Chinese Governance 2(4):411–436. https://doi.org/10.1080/23812346.2017.1379166
40. Steuer B (2020) Governing China's Informal Waste Collectors under Xi Jinping: Aligning Interests to Yield Effective Outcomes. Journal für Entwicklungspolitik XXXVI (1-2020):61–87.

41. Steuer B (2021) Identifying effective institutions for China's circular economy: bottom-up evidence from waste management. Waste Manage Res 39(7):937–946. https://doi.org/10.1177/0734242X20972796
42. Tong D, Wu Y, MacLachlan I, Zhu J (2021) The role of social capital in the collective-led development of urbanising villages in China: The case of Shenzhen. Urban Studies 58(16):3335–3353. https://doi.org/10.1177/0042098021993353
43. Tong X, Tao D (2016) The rise and fall of a "waste city" in the construction of an "urban circular economic system": The changing landscape of waste in Beijing. Resour Conserv Recycl 107:10–17. https://doi.org/10.1016/j.resconrec.2015.12.003
44. White book on China WEEE recovery and comprehensive reutilisation businesses (2018) China Household Electronic Appliances Research Institute (CHEARI). http://img.cheerue.com/D5C994E3-F0CF-4E81-676E-B20D75A511B0_thinkv_2018-05-16_5afbdc128602f.pdf
45. Zacho KO, Mosgaard M, Riisgaard H (2018) Capturing uncaptured values—a Danish case study on municipal preparation for reuse and recycling of waste. Resour Conserv Recycl 136:297–305. https://doi.org/10.1016/j.resconrec.2018.04.031

Open Access This chapter is licensed under the terms of the Creative Commons Attribution 4.0 International License (http://creativecommons.org/licenses/by/4.0/), which permits use, sharing, adaptation, distribution and reproduction in any medium or format, as long as you give appropriate credit to the original author(s) and the source, provide a link to the Creative Commons license and indicate if changes were made.

The images or other third party material in this chapter are included in the chapter's Creative Commons license, unless indicated otherwise in a credit line to the material. If material is not included in the chapter's Creative Commons license and your intended use is not permitted by statutory regulation or exceeds the permitted use, you will need to obtain permission directly from the copyright holder.

Conclusions

Daan Schraven, Martin de Jong, Zhaowen Liu, and Xin Tong

Abstract This conclusion chapter synthesizes the contributions from the preceding chapters to address the central question of how a city can govern its Urban Waste Management System (UWMS) to achieve circularity and inclusion. By integrating insights from diverse perspectives, we highlight the key principles and strategies that enable cities to transition towards more sustainable and inclusive waste management practices. The synthetic UWMS framework introduced earlier in the book serves as a comprehensive tool for understanding the complex dynamics of circularity and inclusion in urban contexts. This chapter discusses the interconnectedness of governance, policy, technology, and community participation, emphasizing that effective UWMS governance requires a multi-layered approach. It concludes with practical recommendations for policymakers and urban planners, outlining pathways for achieving a circular economy while fostering social equity and inclusion within urban waste systems. Ultimately, it calls for an adaptive, collaborative governance model to ensure long-term sustainability and inclusivity in urban waste management.

Keywords Urban waste management system · CE policies · Transformation

---

D. Schraven (✉)
Department of Management in the Built Environment, Delft University of Technology, Delft, The Netherlands
e-mail: d.f.j.schraven@tudelft.nl

M. de Jong
Rotterdam School of Management & Erasmus School of Law, Erasmus University Rotterdam, Rotterdam, The Netherlands

Institute for Global Public Policy, Fudan University, Shanghai, China

Smart City Institute, HEC Liège, University of Liège, Liege, Belgium

M. de Jong
e-mail: w.m.jong@law.eur.nl

Z. Liu
Delft University of Technology, Delft, The Netherlands

X. Tong
College of Urban and Environmental Sciences, Peking University, Beijing, China
e-mail: tongxin@urban.pku.edu.cn

Informal waste pickers or recyclers are often invisible in a city's busy streets. But if one thing has become clear from the various chapters in this book: waste management practices happen organically in all parts of the city, and the informal activities form an integral part of the waste management system, as well as of its aspirations to higher levels of inclusion and circularity. In the theoretical and empirical parts of the book, we took a deep dive into the concepts of inclusion and circularity in urban waste management systems, coming from both the academic literature and real-life cases in the Netherlands and China. With a bird eye's view, the chapters sought to address the main question introduced in the introduction of the book: *How should a city govern the entirety of its urban waste management system such that it achieves inclusive and circular aspirations?*

Each chapter provided an answer to this question and elucidated specific lessons, experiences, and insights. Yet, in a way, a succinct answer to this question helped us synthesize all these individual stories and reviews into a more integrated approach that a city should seek to set up if it wishes to plan its urban waste management system or make changes to its existing system in place.

1 Synthetic Framework for UWMS for Circular Economy Transformation

For this synthesis to be possible, it is important to first sketch the layers of the urban waste management system that have been featured in the topics and cases of the parts A, B, and C. Figure 1 shows these layers as the:

i. properties of separate UWMS components (from Chapter 1);
ii. schemata of boundaries, actors, flows, and indicators of an UWMS (from Chapter 4); and
iii. process methodology for designing inclusive and circular UWMS (from Chapter 4).

First, the smallest layer of the UWMS involves the properties of each separate node in the overall system (i). As adapted from Liu et al. [10], Chapter 1 described that this building block involves a physical facility and set of actors that are involved in the facility, for example through operating or supplying the facility. For example, in a sorting center (physical facility), waste is brought by a transporter and then separated by employees (actors). These actions organize the logistics of the waste travelling through the UWMS from node to node.

Second, multiple nodes make up the entire UWMS (ii). In this system, the waste logistics are managed through the dynamics between the components. Yet, for the UWMS to function, Chapter 4 pointed to the importance of the urban context in which an UWMS operates. This includes the description of the system boundaries, actors, flows and indicators. For example, a city could wish to collect all the waste in its city boundaries (boundaries) and would have several collection points (actors)

Conclusions

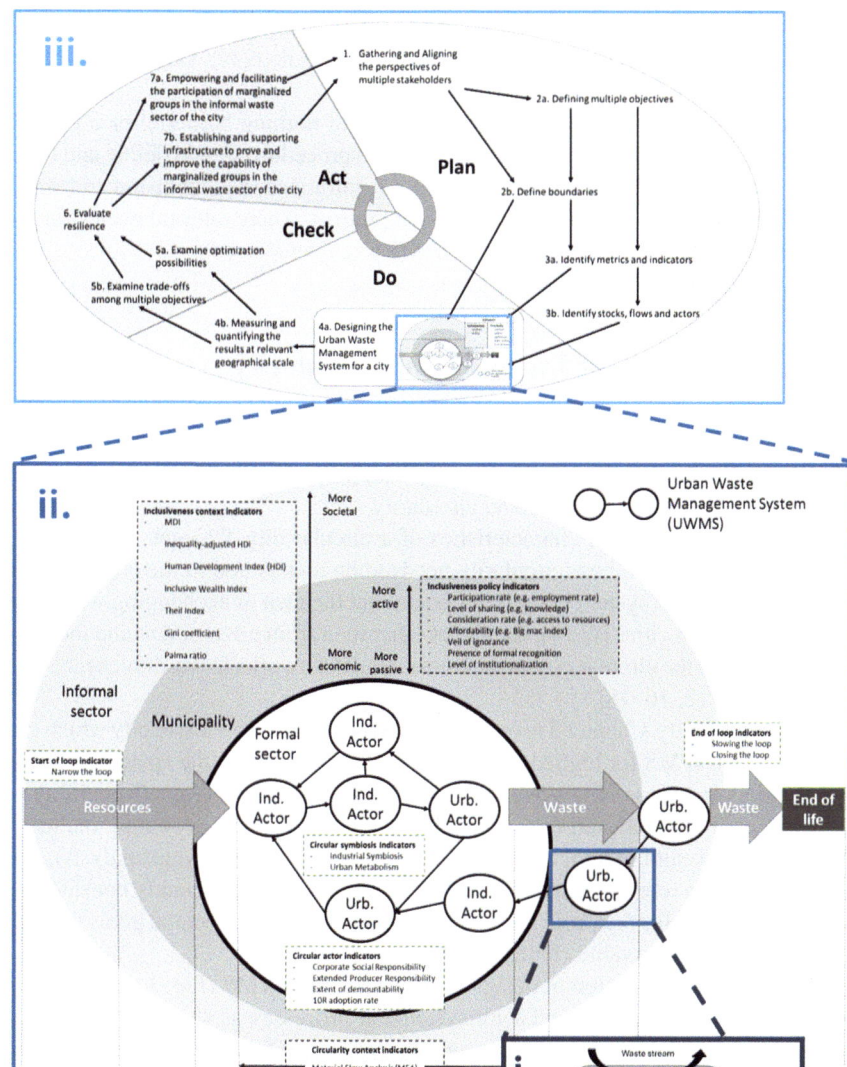

Fig. 1 Layers of an UWMS

where it separates municipal solid waste by paper, plastics, and biowaste (flows) and measures the extent that this is recycled and reused (indicators). Details of all these aspects make up the UWMS design.

Third, a city does not develop its UWMS out of nothing but requires a process methodology (iii). Chapter 4 describes that such a procedure should define and redefine the UWMS, its function for the city and its aspirations (e.g. inclusion and circularity). The process is introduced as a cyclical process, where relevant events for the city ignite mechanisms that attune the UWMS to citizens' needs.

2 Design UWMS for Inclusion and Circularity

The theoretical underpinnings in part A did not only help to create a framework for the UWMS (Chapter 4), but also concluded with propositions as to how the UWMS should be designed for inclusion and circularity.

With the definition and characteristics of a circular city, Chapter 2 established several cornerstones on how circularity needs to be understood in the context of an UWMS. As the hard system boundaries contradict the idea of exchanging waste and resources between cities [2, 14], the actors other than from government and industry filled the gap for the circular city in the transnational networks for resources recovery and recycling [12, 16–18].

More specifically, Chapter 3 revealed six dimensions of an inclusive city which can be tied to the UWMS for higher circularity (Fig. 1, ii), namely: the spatial inclusion (e.g. the sites and areas of UWMS in the city), social inclusion (e.g. affordability of services), political inclusion (e.g. level of institutionalization), economic inclusion (e.g. inclusive wealth index), environmental inclusion (e.g. life cycle analysis), and cultural inclusion (e.g. participation rate). It is important to deliberately consider the informal activities for their role and effectiveness in achieving societal goals through reshaping the circular value chains [9, 13].

Besides the identification of design elements of the UWMS (Fig. 1, ii), inclusion and circularity can be designed with systematic processes as shown in Chapter 4 (Fig. 1, iii). With operational monitoring techniques, five types of circularity indicators were identified: actor indicators (e.g. extended producer responsibility), context indicators (e.g. material flow analysis at city scale), symbiosis indicators (e.g. urban metabolism), start of loop and end of loop indicators (e.g. narrowing and slowing the loop), combined with 2 inclusion indicators: policy indicators (e.g. level of sharing) and context indicators (e.g. Theil index). If an UWMS deploys these circularity and inclusion aspects effectively, it can make the UWMS regenerative for urban prosperity and solidarity [1, 4, 7, 15].

3 Building Links Between Locales in a Separated World

The empirical insights from policy to practices (part B) did not only provide practical lessons for policy makers within specific case-contexts, but also offered more generic insights on various layers of the UWMS (Fig. 1, iii).

The transition of an UWMS into a circular and more resilient system is a challenge both locally and globally. Chapter 5 and 6 included an overview over the history of waste management in both the Dutch and Chinese contexts. The local evolutions generally take shape as a cyclical process reaching activities at different levels (Fig. 1, iii).

Compared to the spontaneous evolution of an UWMS in Dutch cities, the Chinese government is more active in issuing new CE policies, with experimental implementation through demonstration projects. A combination of hierarchical control measures and innovation-pushes (pilots and demonstration projects) characterize the circular city initiatives in China, while North American and European governments typically couple innovation policy with economic policy [5, 8]. These combinatorial policy structures show the direct and indirect effects that the CE policies have on the local UWMS contexts, which have a different position in the transnational recycling chains.

The development of an UWMS in Beijing and its relationship with changes in the informal recycling sector since 1970s illustrated the transformation process in urban development. Chapter 7 demonstrated how direct changes were more quickly implemented, but that although indirect effects took longer to generate impact, these eventually overhauled direct policy intentions when this impact did emerge (Fig. 1, iii). The story of Beijing showcased the tensions between a top-down policy to become a zero-waste city and the migrant recyclers that struggle to survive in a city that they work for. It provided evidence of the undeniable importance of social inclusion in the circular economy at the city level. The chapter stipulated that decisions should: (a) go beyond direct changes, like physical deployment of recycling or disposal facilities; (b) collaborate between authorities and verbal and non-verbal stakeholders under mutual understanding; and (c) set it as a community-based system where verbal and non-verbal stakeholders are designed to have a share.

4 Empirical Insights from Practices to Policies

The empirical insights from practices to policies (part C) did not only provide practical lessons for policy makers within specific case-contexts, but also highlighted more generic insights from various layers of the UWMS (Fig. 1, i).

The Waste Journey, an innovative tool demonstrated in Chapter 8, shows the method to capture the waste management practices in a circular economy specifically incorporating the social dimension. The method allowed the authors to identify barriers and opportunities from the stakeholder interactions in waste handling, and finding their motivations for making decisions, overseeing its journey implications with both quantitative and qualitative data about waste management at the smallest scale (Fig. 1, i).

Chapter 9 focused on the practice of post-consumer recycling and the emerging business models in this realm facilitated by Chinese players through internet technology. The chapter shows how the CE policy of Extended Producer Responsibilities (EPR) resulted in the indirect change of emergent business models in the recycling sector as a response to government interventions [6, 11]. It turned out that practices may respond to policies in unanticipated ways, and thereby made a case for proper inclusion strategies in CE policies, also to learn about potential responses and closing the gap between the governments and informal sector stakeholders [3, 19]. Currently, new business models still rely on government subsidies. The diverse ways in which new business models connect the EPR system to production chains and the UWMS (Fig. 1, ii), sometimes create new actors in that system (Fig. 1, i).

Including informal sector practices adds more value to CE interventions than just technically oriented CE policies. Chapter 10 took the urban village of Gangxia in Shenzhen, China, as a case study, where a small-scaled recovery sector of air conditioners co-exists with formal enterprises. The informal recovery activities enhance the network strength, circular capabilities, and service value for the local low-income groups. The community developed quite effective business continuity and pursued circular practices although national or local policies largely ignored their practices, showing that better inclusion of human relationships in formal regulatory requirements could vastly improve the contribution of indirect changes to directed intentions in CE policies in UWMS (Fig. 1, iii).

5 In Sum

How should a city govern its UWMS to match its inclusive and circular aspirations? Fig. 1 offers an initial layered connection between CE policies (Fig. 1, iii), the designed UWMS (Fig. 1, ii) and the individual waste management practices (Fig. 1, i). Figure 2 shows how each chapter contributes to answering this key question with case materials elaborating the synthetic framework for UWMS (Fig. 1). Below we explain the linkages between the framework parts and the chapter contributions.

Figure 1 (i) represents the small-scale individual acting parts of the overall UWMS. Chapter 9 showed that new policies (EPR) can create new actors that operate as functional links within the UWMS. Yet, it also underscores the importance of livelihoods and formal recognition for stakeholders that populate the informal sector. Chapter 10 even highlighted the existence of informal sector practices that were unconnected with the UWMS. Being linked up basically refers to the recognition, stimulation and protection of informal sector practices by official policies.

How can a city pay heed to and learn from the practices in its intended intervening landscape of activities? On this note, Chapter 8 contributed the Waste Journey method, which offered methods of inquiry to follow waste along its management

Conclusions

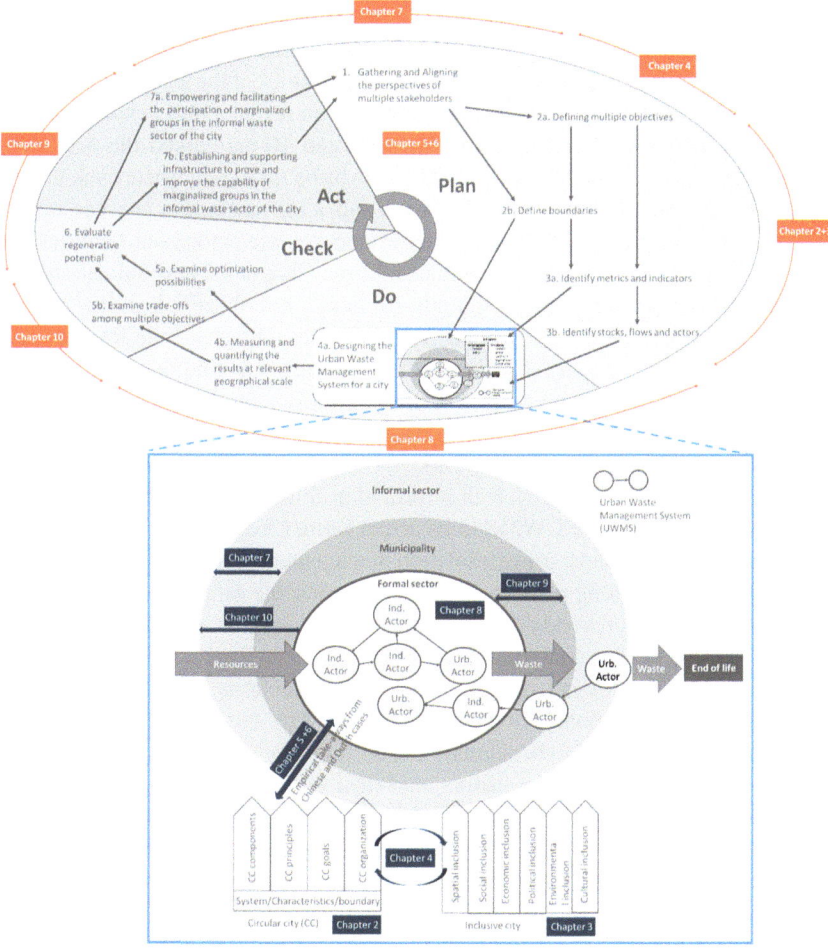

Fig. 2 Mapping of chapter relationships throughout the book

handling path by actors and learning about their motivations to perform their function in the chain. The method demonstrates what makes individual actors tick, and how their actions either stall or facilitate the overall achievement of circular or inclusive aspirations in the UWMS. An important consideration here is the importance of the enduring livelihood needs of the actors, as well as the requirements for the maintenance and upkeep of their human contributions to the inclusive and circular UWMS system. Sketching business models and the role government action plays in securing them can be a means to inform policy makers of relevant needs at the level of waste handling practices, how actors are and can be preserved, boosted or empowered as contributing members of the UWMS.

Figure 1 (ii) shows how multiple individual actors make up the city's overall UWMS. It also shows the role of the city, and the indicators for certain circularity and inclusion targets that it can direct its policy toward. Tying the various roles of the various chapters in the book together, Chapter 8 offered a means to analyze the middle of the UWMS system and its underlying dynamics of the actors and facilities. At this layer, data flowing from such an analysis can reveal a null benchmark from where improvements can be made. Chapters 9 and 10 earmarked the actors and facilities that function as curbing mechanisms to bring discarded waste back into the UWMS for another chance at life. Chapter 6 and 7 showed how a city can expand its inclusion of informal sectors, and thereby incorporate actors and facilities into its policy making that support the circular lifeline. Chapters 5 and 6 illustrated the power-fields that are imposed on the city to dictate how local CE policy making should adopt certain national CE policy requirements. The challenge for cities was mainly in the realm of making effective pushback to national blind spots that overlook contributing members in the informal sector realm. Chapter 2, 3 and 4 underscored how the philosophies of circular and inclusive cities can act as a guide for this, if merged into an inclusive circular UWMS.

Figure 1 (iii) represents the cycle of information collection and decision making at the urban scale shaping the UWMS in its actual form. Chapter 5 illustrated how this operates in the Netherlands, using transition management, how the plan to check the cycle may occur over long periods. In other countries, the same line of thought will lead to different structures and different recommendations for reconfiguring elements and relations. On the Plan component of the cycle, Chapter 2 and 3 offer ways to think about the goals of the city in terms of inclusion and circularity, and how these can be pursued. On the Do component of the cycle, Chapter 8 reveals how the Waste Journey method could be deployed to benchmark the current UWMS by following the waste movements and discovering key actors, and then by analyzing the achievement of those activities to the policy goals. On the Check component of the cycle, Chapter 10 points out how evaluation of the performance of the system in place should recognize the challenges that occur at the individual components, and how the evaluation needs to be reflexive rather than simply following the measurement of outlined indicators only, at least if it is really to evaluate the informal parts of the UWMS. On the Act component of the cycle, Chapter 9 reveals in the case study how the participation of informal actors in the formal regulatory framework should inform changes to a next cycle of policy making to increase the effectiveness of these actors in enhancing their levels of inclusivity and circularity. Chapter 7 reinforces the cyclicality of the framework, with suggestions as to how the informal sector can be made part of such a next cycle, in its long-term case study of the informal recycling sector in Beijing.

Cities have ever-growing footprints in resource consumption burdening the planet. It calls for the transformation towards a circular city worldwide with actions at different locales. The transnational flows of products continue in the end-of-life stage for reuse, recycling and recovery, which complicate the circular value chain in the global production and consumption network. The inclusive circularity is not only locally embedded, but also globally oriented.

Although the modelling approach developed and presented in this book and summarized in this concluding chapter is general in nature and potentially universal in its applicability, it needs to be practically fleshed out and specified for each individual city case. Policymakers and analysts in different urban contexts deal with different types and compositions of waste flows, face different practical problems, have different actor constellations and are embedded within different institutional structures and contractual arrangements. The result is that they can apply the conceptual modelling approach, but each must go through it 'with their own specific data input'. They can also draw lessons from each other provided they do not slavishly copy each other's policies but examine which policy solutions work elsewhere, how their own situation diverges from their example and then account for those differences in their adoption of new goals and instruments.

References

1. Ardi R, Leisten R (2016) Assessing the role of informal sector in WEEE management systems: a system dynamics approach. Waste Manage 57:3–16. https://doi.org/10.1016/j.wasman.2015.11.038
2. Campbell-Johnston K, ten Cate J, Elfering-Petrovic M, Gupta J (2019) City level circular transitions: barriers and limits in Amsterdam, Utrecht and The Hague. J Clean Prod 235:1232–1239. https://doi.org/10.1016/j.jclepro.2019.06.106
3. Chi X, Streicher-Porte M, Wang MYL, Reuter MA (2011) Informal electronic waste recycling: a sector review with special focus on China. Waste Manag 31:731–742
4. Goodwin Brown E, Sosa L, Schröder A, Bachus K, Bozkurt Ö (2020) The social economy—a means for inclusive and decent work in the circular economy? https://www.circle-economy.com/resources/the-social-economy-a-means-for-inclusive-decent-work-in-the-circular-economy
5. Goulder LH, Parry IW (2008) Instrument choice in environmental policy. Rev Environ Econ Policy 2(2):152–174
6. Guide D, Van Wassenhove L (2009) The evolution of closed-loop supply chain research. Oper Res 57:10–18
7. Gutberlet J, Carenzo S (2020) Waste pickers at the heart of the circular economy: a perspective of inclusive recycling from the Global South. Worldw Waste: J Interdiscip Stud 3(1):6. https://doi.org/10.5334/wwwj.50
8. Halpern C (2010) Governing despite its instruments? Instrumentation in EU environmental policy. West Eur Polit 33(1):39–57. https://doi.org/10.1080/01402380903354064
9. Laquian AA, Tewari V, Hanley LM (eds) (2007) The inclusive city: infrastructure and public services for the urban poor in Asia. Woodrow Wilson Center Press/Johns Hopkins University Press [Baltimore]
10. Liu Z, Schraven D, de Jong M, Hertogh M (2023) Unlocking system transitions for municipal solid waste infrastructure: a model for mapping interdependencies in a local context. Resour Conserv Recycl 198:107180
11. Manomaivibool P (2009) Extended producer responsibility in a non-OECD context: the management of waste electrical and electronic equipment in India. Resour Conserv Recycl 53:136–144
12. Paoli F, Pirlone F, Spadaro I (2022) Indicators for the circular city: a review and a proposal. Sustainability 14(19):11848
13. Pokhrel N (ed) (2019) Transforming Kolkata: a partnership for a more sustainable, inclusive, and resilient city. Asian Development Bank, India
14. Stahel WR, Clift R (2016) Stocks and flows in the performance economy. Tak Stock Ind Ecol, 137–158. https://doi.org/10.1007/978-3-319-20571-7_7

15. Tong X, Tao D (2016) The rise and fall of a "waste city" in the construction of an "urban circular economic system": The changing landscape of waste in Beijing. Resour Conserv Recycl 107:10–17. https://doi.org/10.1016/j.resconrec.2015.12.003
16. Tsui T, Derumigny A, Peck D, Van Timmeren A, Wandl A (2022) Spatial clustering of waste reuse in a circular economy. Front Built Environ 154
17. Vanhuyse F, Fejzić E, Ddiba D, Henrysson M (2021) The lack of social impact considerations in transitioning towards urban circular economies: a scoping review. Sustain Cities Soc 75:103394
18. Williams J (2021) Circular cities: what are the benefits of circular development? Sustainability 13(10):5725
19. Zeng X, Duan H, Wang F, Li J (2017) Examining environmental management of e-waste: China's experience and lessons. Renew Sustain Energy Rev 72:1076–1082

Open Access This chapter is licensed under the terms of the Creative Commons Attribution 4.0 International License (http://creativecommons.org/licenses/by/4.0/), which permits use, sharing, adaptation, distribution and reproduction in any medium or format, as long as you give appropriate credit to the original author(s) and the source, provide a link to the Creative Commons license and indicate if changes were made.

The images or other third party material in this chapter are included in the chapter's Creative Commons license, unless indicated otherwise in a credit line to the material. If material is not included in the chapter's Creative Commons license and your intended use is not permitted by statutory regulation or exceeds the permitted use, you will need to obtain permission directly from the copyright holder.

MIX
Papier aus verantwortungsvollen Quellen
Paper from responsible sources
FSC® C105338

If you have any concerns about our products,
you can contact us on
ProductSafety@springernature.com

In case Publisher is established outside the EU,
the EU authorized representative is:
**Springer Nature Customer Service Center GmbH
Europaplatz 3, 69115 Heidelberg, Germany**

Printed by Libri Plureos GmbH
in Hamburg, Germany